Understanding Electricity and Electronics Principles

Written by: Training & Retraining, Inc.
Revised by: David L. Heiserman

Howard W. Sams & Co.
A Division of Macmillan, Inc.
4300 West 62nd Street, Indianapolis, IN 46268 USA

©1987 by Howard W. Sams & Co.

FIRST EDITION
FIRST PRINTING—1986

International Standard Book Number: 0-672-27061-7
Library of Congress Catalog Card Number: 86-63109

Acquisitions Editor: Greg Michael
Editor: Madelaine Cooke
Production Art: Jeanie Bonecutter
Cover Art: Diebold Glascock Advertising, Inc.
Cover Photography: Cassell Productions, Inc.
Components Courtesy of: Warren Radio
Compositor: Shepard Poorman Communications Corp.

Printed in the United States of America

Table of Contents

Preface

Few modern technologies have exerted as much influence on our society as has electronics. For the past hundred years the world has been changed radically by developments in electricity and electronics. First there was the light bulb; then came electric motors, generators, the refinement of vacuum tubes, and radio. After World War II, we were introduced to television and primitive types of computers. We are now entering the "information age," supported by efficient and economical computers.

There is always more to learn and more to do. If you are a newcomer to the world of electronic technology, this book is for you. The main objective of the book is to introduce you to basic principles and to show you how they apply to the electronic devices that are becoming familiar parts of everyday living.

Like other books in this series, this book builds understanding through the subject in a step-by-step manner. Knowledge and confidence are gained as each subject is completed.

Each chapter concludes with a chapter summary, a list of key words presented in the chapter, and a multiple-choice quiz. Study the summary carefully. If any of the ideas in the summary seem different from what you recall, review the text material.

The list of key words reviews terms introduced in the chapter. If you are uncertain about the meaning of any of the terms, review the relevant portions of text and consult the glossary.

The end-of-chapter quiz will test your understanding of important technical ideas and mathematical procedures. Consider all the choices carefully; only one choice is correct, but many of the other choices are based on common errors or misunderstandings of the topic. You can check your responses to the questions by consulting the list of answers in the back of the book.

This book has been carefully designed to make your learning process as simple and meaningful as possible. You must expend the effort necessary for translating the material into ideas that you can remember and use.

D.L.H.

The World of Electricity and Electronics

ABOUT THIS CHAPTER

You are about to become acquainted with the fascinating world of electricity and electronics. You are going to learn what electricity is, what it does, and how it does it. You will use this information to gain a better understanding of what electrical devices are all about, how they work, and how to test and repair them.

ELECTRICITY

Voltage is the force that causes current to flow.

Electricity is a combination of a force called "voltage" and the movement of invisible particles known as "current." The force of voltage can be compared to the force generated by a water pump, which moves water through a distribution system, generally an arrangement of pipes. Voltage is the force that causes current to flow through a system of wires.

Current—the movement of invisible particles—causes electrical devices to operate. We cannot see current, but we can determine its presence by the effects it produces. *Figure 1-1*, for example, shows how the voltage force from a battery causes electrical current to flow through wires and an electrical motor. The current is invisible, but it produces the effect of making the motor run. Current flows through the wires of an electrical device much the same as water flows through pipes.

Figure 1-1.
The Effect of Current

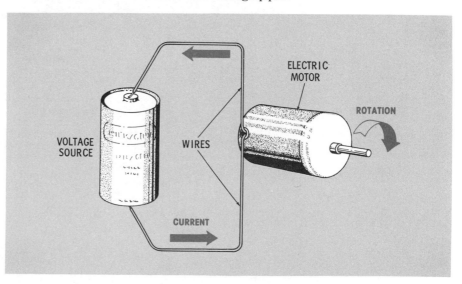

VOLTAGE SOURCE

WIRES

CURRENT

ELECTRIC MOTOR

ROTATION

Current is a flow of invisible particles called electrons that causes an electrical device to operate. Current cannot flow through a broken wire.

Current actually consists of invisible atomic particles called "electrons." So voltage is a force that causes current, in the form of electrons, to move through wires and electrical devices.

There is one important difference between current in wires and water in pipes: water can flow out of a broken pipe, but current cannot flow out of a broken wire. In fact, current cannot flow anywhere in a broken wire. When a wire is broken, the force of the voltage is removed from the motor, as shown in *Figure 1-2*. The circulating pump in the working system creates a force that moves hot water through the pipes and radiator. The battery creates a force that moves current through the wires and causes the motor to run. The wire and pipe are broken open in the broken systems. In these instances, the circulating pump forces water to flow out of the pipe, but even though the battery still creates a voltage force, current does not flow out of the wire.

Figure 1-2.
Working Systems and
Broken Systems

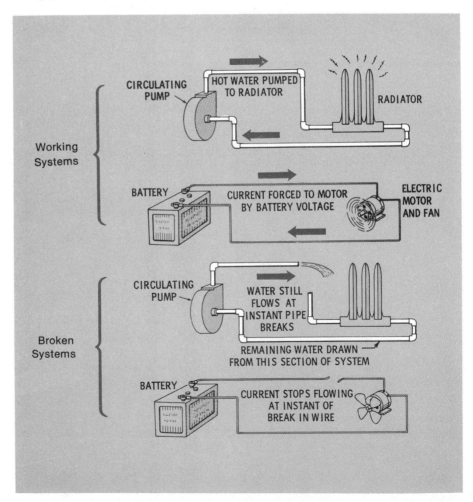

SOME BASIC ELECTRICAL DEVICES

Every electrical device uses voltage and current to produce at least one kind of useful effect. Heat and magnetism are examples of such effects.

Heating Effects

Current flowing through a wire can produce useful heating effects.

Under certain conditions, current flowing through wires can create noticeable amounts of heat. The amount of heat given off by such wires is determined by the type of metal in the wire, the size of the wire, and the amount of current forced to flow through it.

High current flow produces more heat in the same size and type of wire than low current flow. If the current is the same, a smaller wire gives off more heat than one that is larger in diameter. Also, some metals produce more heat than others as the result of current flowing in them.

In fact, manufacturers select the size or type of wire that will produce a desired amount of heat. To do this, they must know the amount of current that will flow through it. *Figure 1-3* shows heat energy given off by a wire that has an electrical current flowing through it.

**Figure 1-3.
Current Flow Heats
Wires**

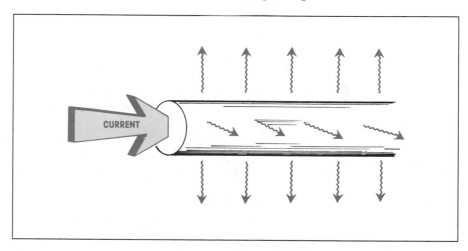

Electrical appliances, such as the iron and toaster shown in *Figure 1-4*, use the heat produced by current flowing through a length of special wire.

**Figure 1-4.
Heating Appliances**

Lighting Effects

A common electric light bulb is not usually regarded as a heating device, but the filament wire in a bulb is heated to a brilliant, white-hot temperature by the current flowing through it. This is illustrated in *Figure 1-5.* A device that uses electrically generated heat to produce useful amounts of light is called an incandescent lamp. Other kinds of lamps, such as fluorescent lamps and light-emitting diodes (LEDs), also produce light, but they do not rely on electrical heating effects.

**Figure 1-5.
Light Production in an
Incandescent Light
Bulb**

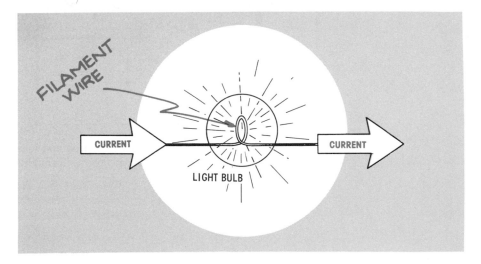

Magnetic Effects

When current flows through a coiled length of wire, the coil acts like a magnet. This can be proven by experimenting with an electromagnet such as the one shown in *Figure 1-6*.

**Figure 1-6.
Production of a
Magnetic Field**

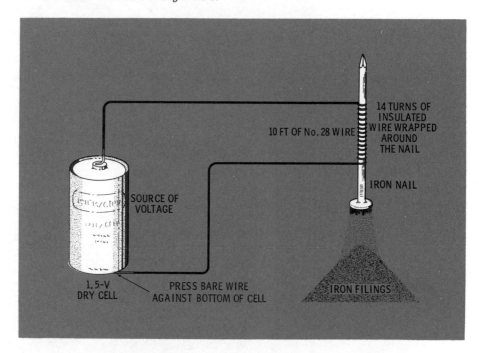

Current flowing through a wire produces a magnetic field.

Current flowing through a wire develops a magnetic field. This field is called an "electromagnetic force" because it is the result of the flow of electric current. If the magnetic field passes through certain kinds of metal, such as soft iron, the metal becomes magnetized and takes on the properties of a magnet.

The electromagnet retains its magnetic capability—continues to attract the iron filings—as long as current flows through the coil of wire. When the current stops, the metal gradually loses its effectiveness as a magnet.

Electric Motors

To produce motion, electric motors use the magnetic fields produced by current flowing through a wire.

An electric motor is actually a device that transforms current flow into mechanical force. *Figure 1-7* shows a common kind of electric motor and the way it uses magnetic fields to change electrical energy into mechanical energy.

**Figure 1-7.
Magnetic Fields in
Electric Motors**

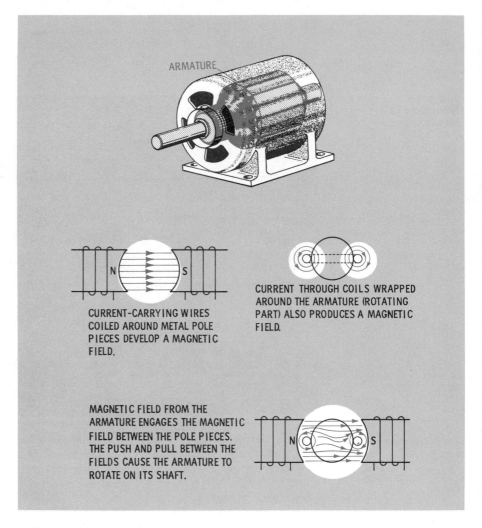

ARMATURE

CURRENT-CARRYING WIRES
COILED AROUND METAL POLE
PIECES DEVELOP A MAGNETIC
FIELD.

CURRENT THROUGH COILS WRAPPED
AROUND THE ARMATURE (ROTATING
PART) ALSO PRODUCES A MAGNETIC
FIELD.

MAGNETIC FIELD FROM THE
ARMATURE ENGAGES THE MAGNETIC
FIELD BETWEEN THE POLE PIECES.
THE PUSH AND PULL BETWEEN THE
FIELDS CAUSE THE ARMATURE TO
ROTATE ON ITS SHAFT.

A Simple Telegraph System

A simple telegraph system, illustrated in *Figure 1-8*, also makes
use of electromagnetic forces. Current flowing through a coil of wire
creates a magnetic force that operates a buzzer or other noise-producing
device. The sounds from the buzzer are the dots and dashes sent by the
operator. When the telegraph key is closed (pressed down), current flows
from the battery through the coil of wire in the buzzer. The resulting
magnetic force causes a movable metal plate to be attracted to the soft-iron
core of the coil, thus producing a clicking or buzzing sound. In this manner,
dots and dashes are transmitted from the telegraph key, through the wire,
and to the buzzer. Common household doorbells operate according to the
same principle.

**Figure 1-8.
A Simple Telegraph
System**

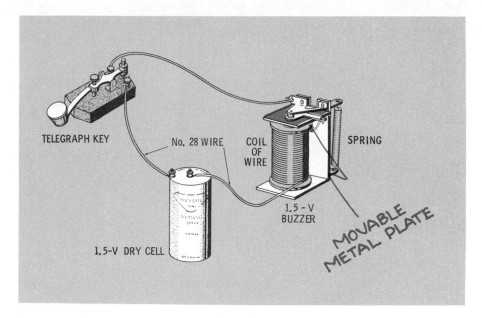

Loudspeakers

The loudspeaker in a radio, television, or other kind of audio device is operated by magnetic forces created by current flowing in a coil of wire.

The loudspeaker shown in *Figure 1-9* consists of a permanent magnet and a coil of wire cemented to a paper cone. Electrical current representing a voice, music, or other kind of sound flows through the coil, creating magnetic forces that cause the coil and cone to be attracted and repelled by the permanent magnet. The movement of the paper cone creates changes in air pressure that are heard as sounds. Telephone earpieces and stereo headsets work in much the same fashion.

SOME PRINCIPLES THAT APPLY TO ALL KINDS OF ELECTRICAL DEVICES

Appliances that require heating effects can use the heat generated by current flowing through heating wires; jobs calling for mechanical motion can be performed by the forces developed by magnetic fields. An electric clothes dryer is an example of an appliance that uses both effects. The heat for drying the clothes is generated by running an electric current through a length of heater wire; the tumbling motion is produced by running an electric current through the electromagnetic coil of a motor.

Figure 1-9.
A Simple Loudspeaker

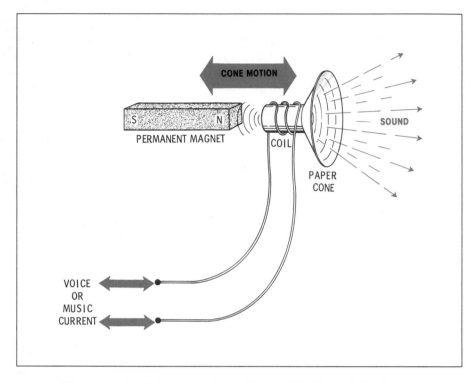

Many complex jobs are performed by combinations of simple devices, that is, two or more simple devices working together to perform a complex task. These devices fall into three categories: input converters, processing devices, and output converters. They work together as shown in *Figure 1-10*.

Figure 1-10.
Three Components of Complex Electrical Systems

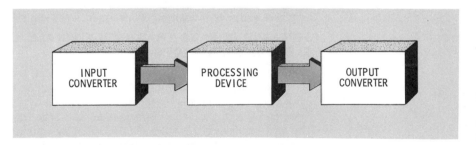

Input Converter

The purpose of an "input converter" in any electrical process is to convert some form of energy, such as sound, light, heat, or pressure, into voltage and current. These forms of energy can be processed by electrical devices after they are converted into the voltage and current forms. Microphones and the keyboards for personal computers are examples of input converters.

Processing Device

A "processing device" changes the amount or form of voltage and current to that required by the output converter. Among the many functions these devices perform are changing small voltages into larger ones, switching and connecting telephone lines, combining many different voltages to control automatic output converters, and producing radio waves.

Output Converter

Very few end results are produced by voltage and currents alone. Therefore, a device is needed that converts voltage or current into radio waves, sound, motion, heat, light, pictures, or other useful forms. This device is called an "output converter."

WHAT HAVE WE LEARNED?

1. Voltage is a force that causes current to flow through a circuit.
2. Current is the flow of invisible particles called electrons.
3. Current cannot flow through a broken wire.
4. Current flowing through wires can produce heat.
5. An incandescent light bulb uses electrically generated heat in the filament to produce light.
6. Current can produce magnetic effects.
7. Electric motors and loudspeakers use the magnetic effects of current flow.
8. Complex electrical circuits can be divided into input converters, processing devices, and output converters.
9. An input converter converts a form of energy into electrical energy.
10. A processing device changes the form or amount of current originally provided by an input converter.
11. An output converter converts electrical energy into some other useful form of energy.

KEY WORDS

Current	Output converter
Electromagnetic force	Processing device
Electrons	Voltage
Input converter	

Quiz For Chapter 1

1. Which one of the following statements best describes voltage?
   a. Voltage is a force that causes current to flow.
   b. Voltage is the flow of invisible particles through a circuit.
   c. Voltage can be compared to the flow of water through a pipe.

2. Which one of the following statements best describes electric current?
   a. Current is the force that causes voltage to flow through a circuit.
   b. Current is the flow of invisible particles through a circuit.
   c. Current can force electrons to flow out of the end of a broken wire.

3. An incandescent light bulb:
   a. converts electrical energy directly to light energy.
   b. converts magnetic energy directly to light energy.
   c. converts electrical energy into heat and heat into light.
   d. converts electrical energy into magnetic energy and magnetic energy into mechanical energy.

4. An electric motor:
   a. converts voltage directly to mechanical energy.
   b. converts electrical energy into magnetic energy and magnetic energy into mechanical energy.
   c. converts electrical energy into heat and heat into mechanical energy.
   d. converts electrical energy into heat and heat into magnetic energy.

5. A loudspeaker is an example of:
   a. an input converter.
   b. a processing device.
   c. an output converter.

6. A microphone is an example of:
   a. an input converter.
   b. a processing device.
   c. an output converter.

7. In a stereo system, the phonograph cartridge is an example of:
   a. an input converter.
   b. a processing device.
   c. an output converter.

8. In a stereo system, the amplifier is an example of:
   a. an input converter.
   b. a processing device.
   c. an output converter.

Basic Electrical Circuits

ABOUT THIS CHAPTER

This chapter describes basic electrical circuits—what they are, what they consist of, and what each device in the circuit does. You will become more familiar with voltage and current and will learn the difference between direct current (dc) and alternating current (ac).

A COMPLETE ELECTRICAL CIRCUIT

A complete electrical circuit provides current a path that returns to its starting point.

A circuit is usually defined as any path that returns to its starting point. In electricity, current makes a complete trip through an "electrical circuit." If the circuit is not complete, current does not flow. A broken wire, loose connector, or a switch in its off position prevents current from flowing.

You have now learned two important facts regarding the flow of current: a voltage source causes current to flow, and a complete circuit allows current to flow.

Voltage is the force that causes current to flow. Current is the flow of electrons through a complete circuit.

A complete circuit must have a voltage source, a device that generates the electrical pressure necessary for making electrons flow through a circuit, thus creating a current. A complete circuit must also have a continuous path for current flow from one side of the voltage source to the other. Even when a voltage source is connected to a circuit, the current will not flow unless the path is continuous.

IMPORTANT PARTS OF ELECTRICAL CIRCUITS

All electrical circuits consist of the basic units shown in *Figure 2-1*. Note the source of voltage and current, the wires for carrying the current, connectors, and the electrical device, which in this example is a lamp, although it could be any electrical device. Many electrical circuits contain more than one device to be operated.

Now that you are familiar with the basic parts of an electrical circuit, you are ready to learn more about each of the parts.

The Voltage Source

Figure 2-2 shows two common sources of voltage and current: a flashlight cell and a wall outlet. A battery, or cell, is a source of voltage and current. An electrical wall outlet is another widely used source of voltage and current. Actually the wall outlet is part of another circuit that has a generator as a voltage source. There may be many miles of wire between the generator and the outlet.

**Figure 2-1.
A Basic Electrical
Circuit**

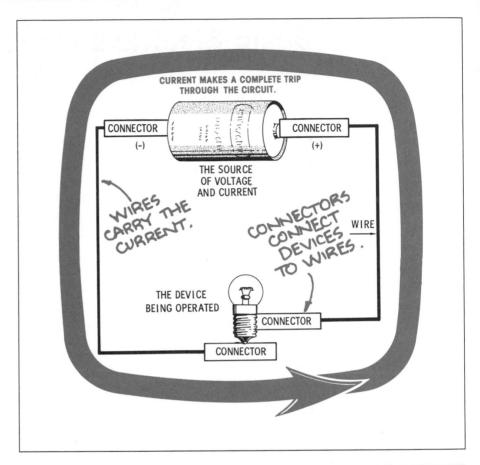

**Figure 2-2.
Common Sources of
Voltage and Current**

SAMS™

Sams books cover a wide range of technical topics. We are always interested in hearing from our readers regarding their informational needs. Please complete this questionnaire and return it to us with your suggestions. We appreciate your comments.

Book Marketing Book

1. Which brand and model of computer do you use?
- ☐ Apple _____
- ☐ Commodore _____
- ☐ IBM _____
- ☐ Other (please specify) _____
_____

2. Where do you use your computer?
- ☐ Home   ☐ Work

3. Are you planning to buy a new computer?
- ☐ Yes   ☐ No
If yes, what brand are you planning to buy? _____

4. Please specify the brand/type of software, operating systems or languages you use.
- ☐ Word Processing _____
- ☐ Spreadsheets _____
- ☐ Data Base Management _____
- ☐ Integrated Software _____
- ☐ Operating Systems _____
- ☐ Computer Languages _____

5. Are you interested in any of the following electronics or technical topics?
- ☐ Amateur radio
- ☐ Antennas and propagation
- ☐ Artificial intelligence/expert systems
- ☐ Audio
- ☐ Data communications/telecommunications
- ☐ Electronic projects
- ☐ Instrumentation and measurements
- ☐ Lasers
- ☐ Power engineering
- ☐ Robotics
- ☐ Satellite receivers

6. Are you interested in servicing and repair of any of the following (please specify)?
- ☐ VCRs _____
- ☐ Compact disc players _____
- ☐ Microwave ovens _____
- ☐ Television _____
- ☐ Computers _____
- ☐ Automotive electronics _____
- ☐ Mobile telephones _____
- ☐ Other _____

7. How many computer or electronics books did you buy in the last year?
- ☐ One or two   ☐ Three or four
- ☐ Five or six   ☐ More than six

8. What is the average price you paid per book?
- ☐ Less than $10   ☐ $10-$15
- ☐ $16-$20 ☐ $21-$25 ☐ $26+

9. What is your occupation?
- ☐ Manager
- ☐ Engineer
- ☐ Technician
- ☐ Programmer/analyst
- ☐ Student
- ☐ Other _____

10. Please specify your educational level.
- ☐ High school
- ☐ Technical school
- ☐ College graduate
- ☐ Postgraduate

11. Are there specific books you would like to see us publish? _____
_____
_____

Comments _____
_____
_____

Name _____
Address _____
City _____
State/Zip _____

27061

Book Markram

BUSINESS REPLY CARD

FIRST CLASS   PERMIT NO. 1076   INDIANAPOLIS, IND.

POSTAGE WILL BE PAID BY ADDRESSEE

HOWARD W. SAMS & CO.
ATTN: Public Relations Department
P.O. BOX 7092
Indianapolis, IN 46206

Book Markram

Voltage is measured in volts (V).

Voltage, or electrical pressure, is measured in "volts," abbreviated V.

A fresh flashlight cell, for example, generates 1.5 V of electrical pressure. The lamps in most flashlights require more than 1.5 V to generate the proper amount of light, which is why you generally find two cells in flashlights, each cell providing 1.5 V, for a total of 3 V. Many kinds of battery-operated devices—portable radios and cassette players, for instance—require even more voltage, perhaps four separate cells arranged in such a way that they provide four times 1.5 V, or 6 V of electrical pressure.

A battery is a voltage source that consists of two or more voltage cells.

Technically, a battery is a voltage source that consists of two or more basic cells. A 9-V battery, for example, is composed of six 1.5-V cells. In common usage, however, even single-cell voltage sources are called batteries.

A common wall outlet provides even more voltage for operating household appliances, approximately 120 V. Certain outlets in the home, particularly those used for operating electric ranges and clothes dryers, require 240-V outlets, and some industrial machinery requires up to 440 V. All sources of voltage and current, including the batteries and wall outlets just described, must have at least two connections.

The source of voltage and current in an electrical circuit is similar to a pump in a water system. The pump provides both the pressure and the water to cause a flow through the water system. A voltage source provides electrical pressure (voltage) and current (the electrical equivalent of water) to cause a flow through an electrical circuit. Like the water pump, the source of voltage and current requires an input connection and an output connection. These connections are shown in *Figure 2-3*.

**Figure 2-3.
Connections for Water
Pumps and Sources of
Voltage and Current**

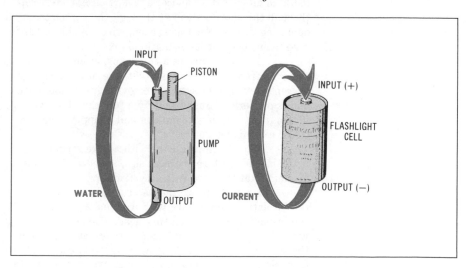

Conductors and Insulators

A good conductor passes electrical current better than a poor conductor does.

Wires provide a path for electric current just as pipes provide a path for water. Metals such as copper and aluminum are most commonly used in the manufacture of electrical wire. Their atomic structures make these metals good "conductors" of current. A good conductor passes electrical current more easily than a poor conductor does. Silver is a better conductor than copper or aluminum but is more expensive. For this reason, most wire is made from copper. When connected into a circuit, a wire is often referred to as a conductor.

An insulator material is a very poor conductor of electric current.

Most non-metals are very poor conductors of electric current. These materials are called "insulators." Plastic is the most commonly used insulator because it can be hard or flexible as required, is easily molded, and can be readily cut when necessary. Because of their better insulating qualities, glass and ceramic materials are used where high-voltage insulators are required.

All electrical wiring uses a good metallic conductor, such as copper or aluminum, to carry current between a voltage source and electrical devices to be operated from it. The metallic conductor in most kinds of wire is covered with an insulating material, usually a soft plastic. The purpose of the insulation is to prevent undesirable electrical contact between conductors.

Wires used for heating purposes are not wrapped with insulation. In these cases, the heating element (wire) is wrapped or formed on an insulating material or supported in air (also a good insulating material) between insulators.

An undesirable contact between two current-carrying conductors is called a short circuit.

If a bare wire comes in contact with another conductor or other metal in an electrical device, a "short circuit" develops. Current thus flows through the short instead of through the complete circuit containing the operating device. For this reason, wire should be handled with sufficient care to ensure that its insulation is not damaged.

To join a wire to a connector, a length of insulation must be removed from the wire. A metal-to-metal connection is required to permit current flow. The process of removing insulation is called "wire stripping" and is properly accomplished by a tool called a "wire stripper." As shown in *Figure 2-4*, both wire cutting and stripping can also be done with a tool called a "diagonal cutter."

When stripping wire with a wire stripper, place the wire in the notch labeled with the correct size, squeeze the handles, and pull the insulation off the wire with the stripper.

Do not squeeze the handles too tightly when attempting to remove the insulation from a wire with diagonal cutters. Just break the surface of the insulation with the cutting head. The cut need not go through to the wire. A steady pull should then part or tear the remaining insulation from the wire. Placing the index finger between the handles prevents the cutters from closing completely and nicking or cutting the wire.

Figure 2-4.
Stripping the Insulation
from Wire

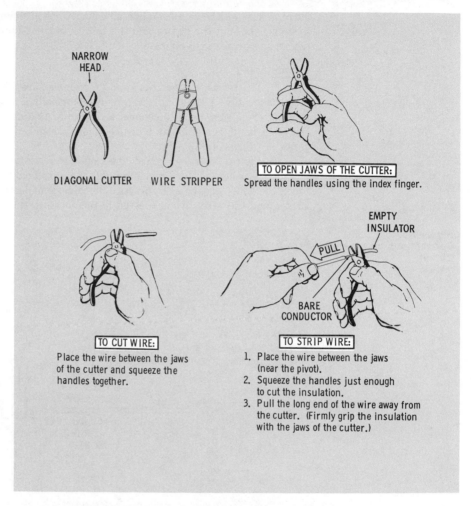

Electrical Devices

Current from a voltage source operates devices such as electric light bulbs, heaters, and motors. Radio and television receivers are also operated by current from voltage sources. These devices process voltage and current contained in the received radio waves by changing the input energy into sound and pictures. The voltage sources make it possible for these devices to perform this process.

As stated earlier, all electrical devices must have two or more connections to a circuit. These connections are used to join conductors to the device, thus completing the circuit and permitting current to flow into and out of the device. The voltage source operates the device by forcing current through the circuit. When all connections are made in the circuit, including those at the device and the source, current flows through the device and causes it to operate.

Connectors

The terms "connectors" and "terminals" are often used interchangeably. A connector, however, is normally a mechanical part, such as a battery clamp, used to connect a conductor to a device. A terminal, on the other hand, is a point on a device where a connection can be made—a screw or other contact point.

Figure 2-5 shows how connections are made to a voltage source and an operating device. The lamp lights when the bare conductors merely touch the lamp terminals. In practice, however, the lamp is placed in a socket, and the wires are connected to the socket terminals.

**Figure 2-5.
Connections for a Lamp
Circuit**

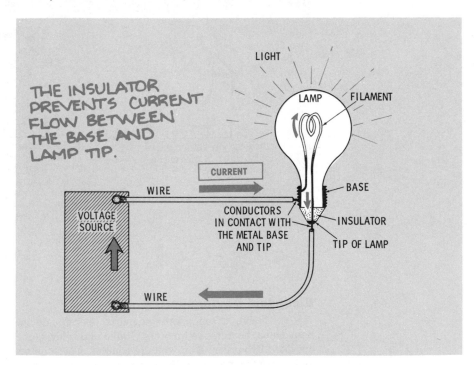

Wires, connectors, and terminals allow current to flow in a circuit because they are made of conducting metals. Care must be taken, however, when joining these parts to each other. After a wire has been stripped, it should be clean of any remaining insulation. Metal at the contact points must be clean and free from the insulating properties of contaminants such as dirt or grease. Sandpaper can be used to clean these junction points when necessary.

When connecting a wire to a terminal, make sure that the screw or clamp makes a tight connection.

A PRACTICAL CIRCUIT

The circuit shown in *Figure 2-6* demonstrates the way all basic circuits are connected. It contains a voltage source, wire, connectors (terminals), and an operating device. The voltage source in this instance is a 1.5-V dry cell. The operating device is a 1.5-V lamp. It is a practical circuit because it has various applications.

Figure 2-6.
Practical Lamp Circuit

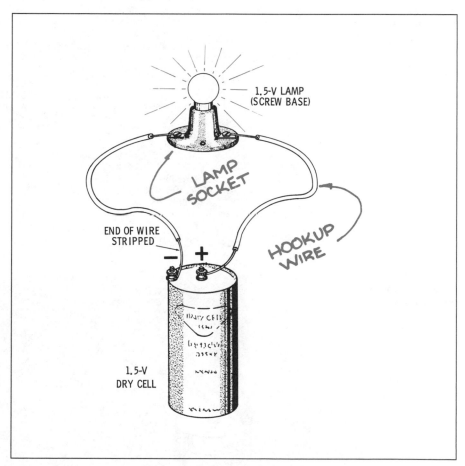

1.5-V LAMP
(SCREW BASE)

LAMP SOCKET

END OF WIRE STRIPPED

HOOKUP WIRE

1.5-V DRY CELL

Current cannot flow through an open circuit. Current flows through a complete, or closed, circuit.

If all the connections are made as shown in the illustration, the lamp will light. If any of the connections are not properly made—a condition known as an "open circuit"—the lamp will not light. An open circuit is a condition that prevents the flow of current. In other words, the circuit is not complete. On the other hand, a "closed circuit" has all of its connections made and forms a complete path for current flow.

SWITCHES

Because it is often desirable to open and close a circuit, nearly all circuits contain some form of switch.

Figure 2-7 shows one simple switch, called a "knife switch" because it has an element resembling the blade of a knife. The basic circuit described above can be reconnected to include a knife switch. Figure 2-8 shows how the connections are made. Be sure you understand what happens to the flow of current when the switch is open and when it is closed.

**Figure 2-7.
Knife Switch**

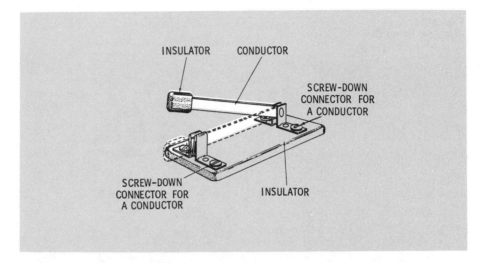

**Figure 2-8.
Practical Lamp Circuit
with Switch**

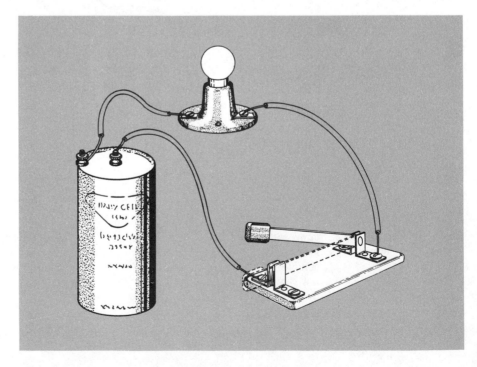

There are many other types of switches, some of which you have used. For example, there are switches on the walls of your home, on appliances, and on the dashboard of your car. Many operate on the knife-switch principle but include springs to provide a positive snap action. Most are enclosed in a sturdy plastic housing.

When a switch is turned to its off position, the contacts are open, and current cannot flow through the switch. When the switch is turned to its on position, the contacts are closed, and current can flow through the switch.

> When a switch is set to its off position, it opens the circuit and does not allow current to flow. When a switch is set to its on position, it closes the circuit and allows current to flow.

VOLTAGE AND CURRENT MEASUREMENT

You have already learned that the volt is the unit of measurement for voltage, or electrical force. The number of volts expresses the amount of electrical force available from the source. Recall that it is this force that causes current to flow. The greater the voltage measurement, the greater the current will be.

> The unit of measurement for current is the ampere (A).

Current is measured in terms of a unit called an "ampere" (sometimes called "amps"). The number of amperes defines the rate of current flow in a circuit. Some flashlight lamps, for example, draw 0.25 ampere (abbreviated 0.25 A) from the voltage source.

A 100-W light bulb draws a little less than 1 A from the 120-V home electrical system. Ten amperes flow through some electric irons, toasters, and heaters. A car battery supplies 100 A or more to an electric starter motor.

Values of voltage and current can be very large or very small. Because it is awkward to talk and write about 500,000 V or 0.003 A, we have developed units that are easier to handle. With this system, the quantities mentioned become 500 kilovolts (kV) and 3 milliamperes (mA), respectively. A kilovolt is 1000 V; a milliamp is 0.001 A.

Table 2-1 will help you convert from one unit to another.

**Table 2-1.
Conversion Table**

When You See	Do This to Convert	Example
mega or M	Multiply by 1,000,000	2 megavolts is 2,000,000 volts
kilo or k	Multiply by 1000	5 kiloamperes is 5000 amps
milli or m	Divide by 1000	7 millivolts is 0.007 volt
micro or μ	Divide by 1,000,000	9 μamperes is 0.000009 ampere
nano or n	Divide by 1,000,000,000	5 nanovolts is 0.000000005 volt
pico or p	Divide by 1,000,000,000,000	5 picoamperes is 0.000000000005 ampere

DIRECT CURRENT

A current that flows only in one direction is called a "direct
current" (abbreviated dc). Dry cells and batteries are common sources of
direct current. Some types of electric generators also supply direct current.
Later you will learn about a power supply that provides direct current for
use within a wide range of electronic devices, including radio and television
receivers, large stereo systems, and personal computers.

A voltage that provides direct current is a direct voltage. Because
direct current is abbreviated dc, the abbreviation is used to identify direct
voltage—dc voltage.

The terminals of a voltage source are marked with plus (+) and
minus (−) signs indicating the direction in which current flows in a circuit.
There are two systems describing the direction of current flow:
"conventional current flow" and "electron current flow."

Conventional Current Flow

The conventional current theory was the first to be developed.
Benjamin Franklin is considered to be its originator, and it is still being
used in many electrical engineering texts. Conventional current is said to
flow from the positive (+) terminal of a voltage source, through the circuit,
and back to the negative (−) terminal.

Electron Current Flow

Of more recent origin, the electron current theory permits a
clearer explanation of how current flows through electronic circuits. For
this reason, the electron current direction of flow is used in this text. This
theory states that current leaves the negative (−) terminal, flows through
the circuit, and returns to the positive (+) terminal of the voltage source.
Figure 2-9 illustrates this electron current theory.

Bearing in mind that voltage and current are invisible and that
theories of electricity were fairly new in Ben Franklin's time, you can
appreciate the fact that he had a 50-50 chance of guessing the appropriate
direction for current flow. More modern findings about the nature of
electrons suggest that Franklin missed his guess. In any event, if you learn
the rules of electron flow, conventional flow should not be confusing. You
will find it easy to mentally reverse the direction of current flow in order
to understand explanations offered in that form.

Voltage supplies the necessary force, but current does all the work
involved in the operation of any electrical device. In any application,
whether a simple lamp or a complex computer system, at least one
continuous path must be provided between the two terminals of a voltage
source before current can flow.

ALTERNATING CURRENT

A current that reverses its direction of flow at regular intervals is
called "alternating current" (abbreviated ac). You might ask, Why should
we have a current that is constantly changing its direction? The reason is
that alternating current has certain features that make it desirable.

Figure 2-9.
Electron Current Theory

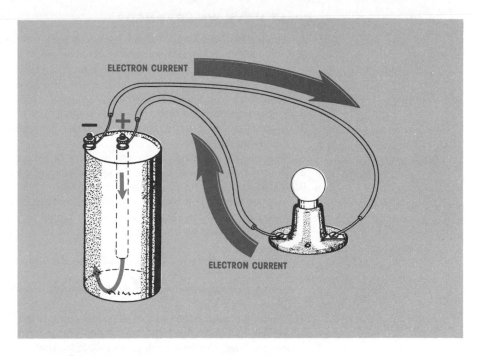

Wall outlets in your home supply an ac voltage. This voltage is produced by generators that may be located many miles away. During the early days of electrical technology, dc was supplied to homes. However, dc can be sent through power lines, or wires, for only short distances before the voltage level drops so low that it is virtually useless.

Ac voltage, on the other hand, can be easily stepped up to a higher value and stepped down to a lower value. This characteristic makes it possible to transmit it economically over long distances—hundreds of miles in some cases. Some power distribution lines carry ac voltages on the order of 400 kV. Power substations located near towns and cities step the voltage down to about 440 V, and the transformers located on neighborhood utility poles step the ac voltage down to 240 V and 120 V for use in individual homes.

The preceding chapter described input converters that convert other forms of energy into voltage and current. Many of these forms, such as sound and radio waves, occur in alternating cycles. Sound waves, for instance, are alternating areas of maximum and minimum air pressure. When these other forms of energy are converted into electricity, as in a telephone mouthpiece, the resulting current is also alternating.

WHAT HAVE WE LEARNED?

1. An electrical circuit provides a complete path for current flow.
2. Every electrical circuit consists of (1) a source of voltage that causes current to flow, (2) conductors that provide a path for the current, (3) electrical devices that are operated by the current, and (4) connectors to join conductors to a source or device.
3. Voltage sources and electrical devices always have at least two connections. All connections must be made in order for current to flow through these sources and devices.
4. Most metals can conduct current and are called conductors.
5. Most non-metals are poor conductors of current and are called insulators.
6. A wire consists of a conductor covered with a plastic insulator material.
7. An open circuit is a condition in which the current path is interrupted. A closed circuit is the same as a complete circuit. A short circuit occurs when a conductor makes an undesirable contact with another conductor or metal part.
8. Switches are designed to open and close circuits and thereby turn a device off and on.
9. The unit of measurement for electrical force, or voltage, is the volt (abbreviated V).
10. The unit of measurement for electrical current is the ampere (abbreviated A).
11. Volt and ampere quantities are often expressed in terms of very large and small values. Prefixes such as mega-, kilo-, milli-, and micro- simplify the expressions.
12. Direct current (dc) always flows in one direction.
13. Electron theory states that current in a dc circuit always flows from the negative (−) terminal of the voltage source, through the circuit, and returns to the positive (+) terminal.
14. Alternating current (ac) reverses its direction of flow at regular intervals.
15. Ac current can be distributed over long distances economically.

KEY WORDS

Alternating current (ac)	Insulator
Ampere	Kilovolt
Battery	Milliampere
Cell	Negative (−)
Closed circuit	Open circuit
Complete circuit	Positive (+)
Conductor	Short circuit
Conventional current flow	Switch
Direct current (dc)	Volt
Electronic current flow	

Quiz For Chapter 2

1. Which one of the following statements best describes conventional current flow?
   a. Current flows in only one direction.
   b. Current flows in two directions at the same time.
   c. Current reverses direction at regular intervals.
   d. Current flows from the negative terminal of a voltage source and returns to the positive terminal.
   e. Current flows from the positive terminal of a voltage source and returns to the negative terminal.

2. Which one of the following statements best describes the electron theory of current flow?
   a. Current flows in only one direction.
   b. Current flows in two directions at the same time.
   c. Current reverses direction at regular intervals.
   d. Current flows from the negative terminal of a voltage source and returns to the positive terminal.
   e. Current flows from the positive terminal of a voltage source and returns to the negative terminal.

3. Which one of the following statements best describes direct current?
   a. Current flows in only one direction.
   b. Current flows in two directions at the same time.
   c. Current reverses direction at regular intervals.

4. Which one of the following statements best describes alternating current?
   a. Current flows in only one direction.
   b. Current flows in two directions at the same time.
   c. Current reverses direction at regular intervals.

5. Which one of the following is a popular source of dc voltage and current?
   a. A common wall outlet.
   b. A flashlight battery.
   c. A flashlight bulb.
   d. Any good conductor.
   e. Any good insulator.

6. Which one of the following is a popular source of ac voltage and current?
   a. A common wall outlet.
   b. A light bulb.
   c. A flashlight bulb.
   d. A cell.
   e. Any good conductor.

7. Which one of the following statements best describes an open circuit?
   a. Current cannot flow through such a circuit.
   b. Current flows easily through the desired path.
   c. Current flows easily, but through an undesirable path.

8. Which one of the following statements best describes a closed circuit?
   a. Current cannot flow through such a circuit.
   b. Current flows easily through the desired path.
   c. Current flows easily, but through an undesirable path.

9. Which one of the following statements best describes a short circuit?
   a. Current cannot flow through such a circuit.
   b. Current flows easily through the desired path.
   c. Current flows easily, but through an undesirable path.

10. Two kilovolts is equal to:
    a. 0.002 V.
    b. 0.2 V.
    c. 200 V.
    d. 2000 V.

11. Thirty-five milliamperes is equal to:
    a. 0.0035 A.
    b. 0.035 A.
    c. 350 A.
    d. 3500 A.

12. Which one of the following statements is true?
    a. A switch set to its off position closes the path for current flow through it.
    b. A switch set to its on position opens the path for current flow through it.
    c. A switch set to its off position completes the circuit through it.
    d. A switch set to its on position completes the path for current flow through it.

How to Use Meters

ABOUT THIS CHAPTER

Because volts and amperes are units of measurement, some device must be used to measure them. Devices used for this purpose are called "meters." You are going to learn about the different types of meters and how to use them to measure voltage and current.

GENERAL PRINCIPLES

Meters convert electrical current into a magnetic field that causes a pointer to move across a scale.

Like motors, meters convert electrical current into mechanical motion. In a motor, magnetic fields caused by current flowing through a coil of wire cause the armature to rotate. Likewise, magnetic fields created by the flow of current through a meter cause a pointer (sometimes called a needle or indicator) to move across a scale. The position where the pointer comes to rest indicates the amount of current flowing through the meter.

All homes and most cars have meters similiar in principle to those discussed in this chapter. One kind of electrical meter measures the consumption of house current. Gasoline, temperature, and other automobile gauges are all basically meters measuring current flow. The quantities being measured are converted into equivalent values of current.

READING THE SCALE ON A METER

Figure 3-1 shows how a current-measuring meter is connected to measure the amount of current flowing in the circuit. Note that the meter is part of the complete circuit between the battery (voltage source) and flashlight bulb (operating device).

Meters are read by noting the position of the pointer on a scale. The pointer sometimes points directly to a number on the scale. More often, however, it points to a division mark between two numbers. When that is the case, the decimal value of the division is added to the lower number.

Scale 1 in *Figure 3-2*, shows the pointer indicating exactly 1 A of current. Scale 2 is indicating 1.5 A, Scale 3 is indicating a current reading of 1.8 A, and Scale 4 shows a current reading of 1.6 A. You are about to learn that you read the scale on other kinds of meters in much the same way.

Figure 3-1.
Current-Reading Meter
in a Circuit

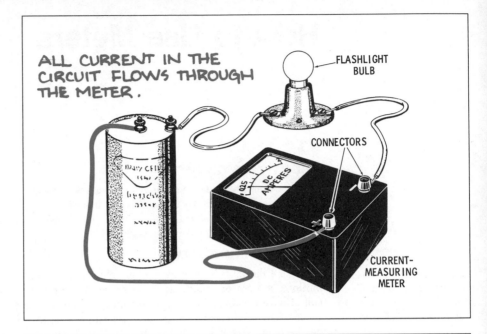

ALL CURRENT IN THE
CIRCUIT FLOWS THROUGH
THE METER.

FLASHLIGHT
BULB

CONNECTORS

CURRENT-
MEASURING
METER

Figure 3-2.
Readings on the Scale
of a Current Meter

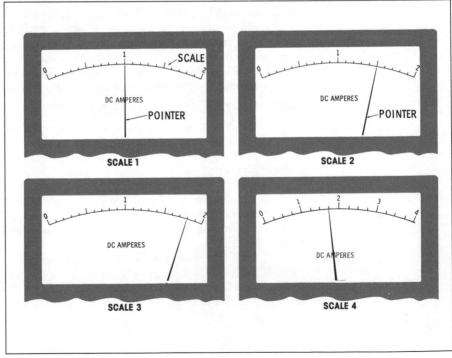

SCALE

POINTER

DC AMPERES

SCALE 1

POINTER

DC AMPERES

SCALE 2

DC AMPERES

SCALE 3

DC AMPERES

SCALE 4

VOLTMETERS

A voltmeter is an instrument used for measuring voltage.

"Voltmeters" are used for measuring voltage. When a voltmeter is connected across the terminals of a voltage source, a small amount of current flows through the meter. The amount of current is proportional to the amount of voltage. The meter scale is graduated (drawn) to give a reading in volts. The procedure for reading a voltmeter scale is similar to that for the current scales you have just read.

Types of Voltmeters

You must use an ac voltmeter for measuring ac voltages. You must use a dc meter for properly measuring dc voltage.

There are two basic types of voltmeters—one for measuring dc voltage and another for ac voltage. Be sure to use the correct one for the type of voltage to be measured. When an ac voltmeter is applied to a dc voltage source, an incorrect measurement will result. But when a dc meter is used for measuring ac voltages, the meter may be damaged. Voltmeters are usually clearly marked as dc or ac meters.

Reading a Voltmeter

As shown in *Figure 3-3*, a voltmeter scale is similar to a current-measuring scale. In this instance, the meter is indicating 34 V. Also note that the meter scale shows you are using a dc voltmeter.

**Figure 3-3.
A Voltmeter Scale**

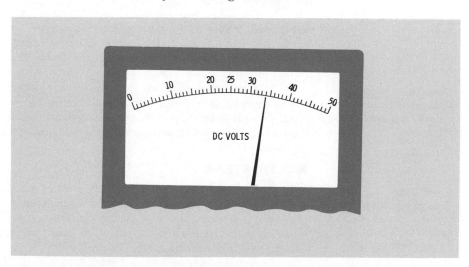

Voltmeters are designed to be read to certain maximum values. From zero to a maximum voltage is called the range of a voltmeter. Some commonly used ranges are 0–10 V, 0–50 V, 0–250 V, and 0–1000 V. The meter in *Figure 3-3* has a range of 0–50 volts.

Always make sure that any voltage to be measured is within the range of the voltmeter you are using. A meter will be damaged if used to measure a voltage greater than the maximum value for which it is designed. Excess voltage causes excess current to flow. As a result, the pointer may be bent in trying to move beyond the end of the scale, or meter circuits may overheat and damage other delicate parts.

AMMETERS

An ammeter is an instrument used for measuring current.

A current-measuring meter is called an "ammeter." It can be used only to measure the amount of current flowing in a circuit. Ammeters used with electrical appliances use current ranges of 0–10 A or 0–30 A. Ammeters used for testing electronic devices most often use much lower ranges: 0–500 μA, 0–10 mA or 0–250 mA.

Using an Ammeter

Always use a dc ammeter for measuring dc current and an ac ammeter for measuring ac current.

Precautions for using ammeters are the same as for using voltmeters. First, you should never use a dc ammeter for measuring ac current, and you should never use an ac ammeter for measuring dc current. Second, you should not attempt to measure a current value that is beyond the range of the meter.

The first precaution can be observed if you know the type of voltage source that is supplying the current. For example, you know that batteries supply dc current (and voltage), and most wall outlets supply ac current (and voltage).

The second precaution can be followed as you gain some practical experience. If your meter has a selection of ranges, always use the highest range first. Then switch to the appropriate lower range to obtain the most accurate reading. Quickly remove the meter from the circuit if the pointer swings beyond the limits of the scale.

Never use a voltmeter to measure current nor an ammeter to measure voltage.

A third precaution should be added to the list: never attempt to use an ammeter to measure voltage nor a voltmeter to measure current. Each meter is designed to measure only certain electrical values. If either type of meter is used for measuring the other kind of value, the meter may be damaged.

MULTIMETERS

A multimeter is a device that can be used for measuring a number of different electrical quantities, including voltage and current.

The term "multimeter" literally means "many meter." A multimeter is a combination voltmeter and ammeter, a single instrument capable of performing many measuring functions. It can be used to measure either ac or dc voltages and currents. The most basic type of multimeter is called a "volt-ohm-milliammeter," or simply VOM.

A multimeter face has a combination of scales that may include several ranges of voltage and current readings. *Figure 3-4* shows a multimeter scale having three ranges. A multimeter also has front-panel switches that allow you to select the kind of reading (ac or dc current or voltage) and the desired range.

**Figure 3-4.
A Typical Multimeter
Scale**

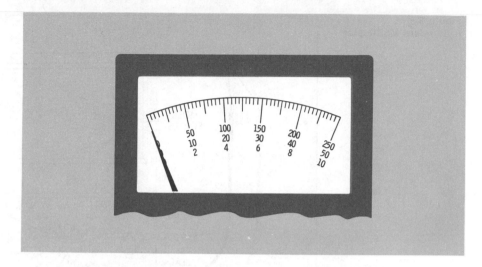

The example in *Figure 3-5* shows the pointer indicating a certain reading. If you happen to be using the 0–10 range, the reading is 7.4. If you are using the 0–50 scale, the reading is 37. And if you have selected the 0–250 scale, the pointer is indicating a reading of 185.

**Figure 3-5.
Reading a Multimeter
Scale**

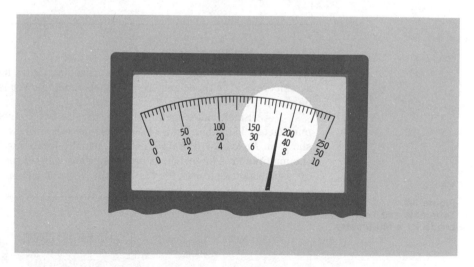

Figure 3-6 shows the front panel of a typical VOM. It may look rather complicated at first, but matters become simpler when you know what units you want to measure and the range you want to use for making the measurements. Once you have that information, you simply set the switches to the desired positions.

**Figure 3-6.
A Standard Multimeter**

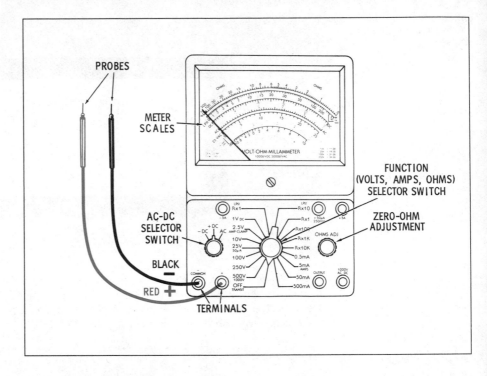

Using a Multimeter

The multimeter shown in the preceding figure measures ac volts, ac amperes, dc volts, dc amperes, decibels (audio power level), and ohms (to be discussed later).

All multimeter measurements require two connections; so a multimeter, like other kinds of electrical devices, has two terminals. As shown in *Figure 3-7*, the terminals are usually located on the front panel of the multimeter. The terminals are sometimes colored red for the positive (+) connection and black for the negative (−) connection.

**Figure 3-7.
Terminals and Test
Leads for a Multimeter**

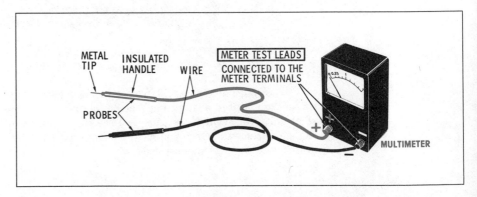

A multimeter requires a pair of "test leads" to connect the meter to the circuit being tested. Test leads are lengths of flexible insulated wire. One end has a means of joining the lead to a terminal on the voltmeter. The other end has a metal probe encased in a sturdy, insulated handle. Like the terminals, the test leads are often red, for the positive connection, and black, for the negative connection.

Measuring Voltage

When voltages are being measured, the probes are touched to the terminals of the voltage source or device. A voltage measurement is always taken across the terminals as shown in *Figure 3-8*. Voltage measurements are never taken between a terminal and an open wire.

**Figure 3-8.
Reading Dc Voltage
with a Multimeter**

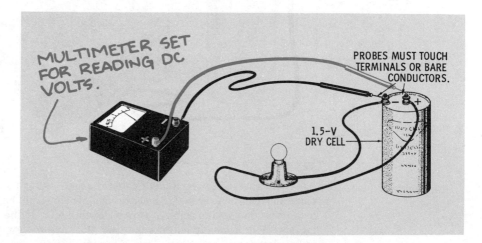

If you are measuring dc voltage levels, you must follow the polarity rule: the polarity marking of the dc voltmeter terminal must be the same as the polarity of the voltage you are measuring. However, the polarity rule does not have to be followed when you are measuring ac voltage levels.

Measuring Current

A multimeter becomes a part of the circuit when measuring current. *Figure 3-9* shows a multimeter properly inserted into a dc circuit. The circuit must be opened (usually at a terminal) and the probes inserted, one on each side of the break.

When measuring dc current, you must follow the polarity rule: dc current should enter the negative terminal of the ammeter and leave by way of its positive terminal. Bearing in mind that the circuit's current flows from the negative terminal of the source and returns to the positive terminal, you should have no trouble following this polarity rule for dc ammeter connections.

Never touch the probes of a multimeter directly across the terminals of a battery when the multimeter is set for a current-measuring function.

**Figure 3-9.
Reading Dc Current
with a Multimeter**

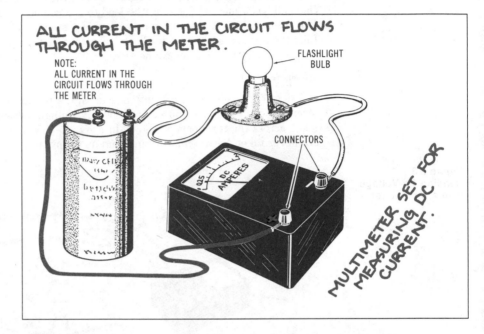

Learning to use a multimeter requires you to think only of the particular function for which you are using the instrument. If you are measuring dc voltage, think of it as a dc voltmeter. If the next measurement is ac current, change your thinking to view the instrument as an ac ammeter. By concentrating on specific applications, you are more certain to make the proper settings and to observe the appropriate measuring precautions.

Digital Multimeters

Recent developments in electronics make it possible to provide digital multimeters at a reasonable cost. Digital multimeters serve the same purpose as standard multimeters but do not use magnetic effects for their operation. Instead of using a pointer and scale to indicate the readings, a digital multimeter shows the readings on a digital display—a display that looks much like the display on an electronic calculator or electronic digital clock. *Figure 3-10* shows the front panel of a typical digital multimeter. One of the advantages of a digital multimeter is that it eliminates the need for reading between the numbers. A digital multimeter automatically displays readings with an accuracy of threee digits or more.

Figure 3-10.
A Digital Multimeter

WHAT HAVE WE LEARNED?

1. Meters are used to indicate the quantity or value of voltages and currents.
2. Meters are read by noting the position of a pointer on a marked scale. Digital meters display the readings as numerals.
3. Voltmeters are used for measuring voltage.
4. Ammeters are used for measuring current.
5. Dc meters should not be used for measuring ac, and ac meters should not be used for dc.
6. The range of a meter is indicated by the highest marking on the scale.
7. Never connect a meter to measure a quantity you know is above the meter's range.
8. A multimeter combines several metering operations. Multimeters typically measure several ranges of ac current and voltage, and dc current and voltage.
9. Voltage measurements are made by connecting a voltmeter's probes across the terminals of the voltage source or device to be measured.
10. When you are making dc voltage measurements, make certain the positive polarity of the circuit is connected to the positive terminal on the meter and that, likewise, the negative polarity of the circuit is connected to the negative terminal on the meter.
11. Current measurements are made by connecting an ammeter into the circuit in a manner that allows the circuit's current to flow through the meter. This normally requires opening the circuit and connecting the meter into it.

12. When you are making dc current measurements, be sure to connect the meter so that the positive lead goes to the more positive point in the circuit, and the negative, to the more negative point.
13. Switches on the front panel of a multimeter let you select the function and scale.
14. When working with a multimeter, forget its many applications and think only in terms of the specific application at hand.

KEY WORDS

Ammeter
Milliammeter
Multimeter
Voltmeter
Volt-Ohm-Milliammeter (VOM)

Quiz For Chapter 3

1. What is the reading on the meter shown in *Figure 3-11*?
   a. 50.
   b. 50.25.
   c. 50.5.
   d. 52.5.
   e. 55.

4. If the meter shown in *Figure 3-11* is an ac voltmeter, what is the largest amount of ac voltage it can measure without possible damage to the meter movement?
   a. 120 mV.
   b. 1.2 V.
   c. 12 V.
   d. 120 V.

**Figure 3-11.
Meter for Quiz
Questions 1–4**

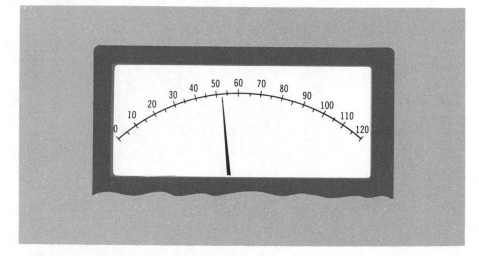

2. What is the range of the meter shown in *Figure 3-11*?
   a. 0–50.
   b. 0–60.
   c. 0–120.
   d. 50–120.
   e. 50–60.

3. If the meter shown in *Figure 3-11* is a milliammeter, what is the largest amount of current it can measure without possible damage to the meter movement?
   a. 120 mA.
   b. 1.2 A.
   c. 12 A.
   d. 120 A.

Figure 3-12.
Meter Connection for
Quiz Question 5

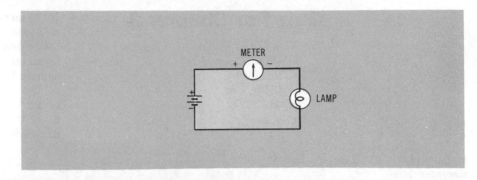

Figure 3-13.
Meter Connection for
Quiz Question 6.

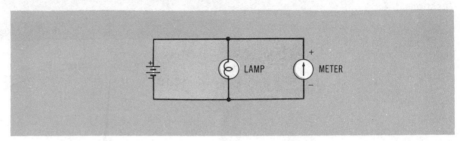

5. The meter shown in the circuit in *Figure 3-12* best demonstrates the proper way to connect:
   a. a voltmeter.
   b. an ammeter.
   c. an ohmmeter.

6. The meter shown in the circuit in *Figure 3-13* best demonstrates the proper way to connect:
   a. a voltmeter.
   b. an ammeter.
   c. an ohmmeter.

7. The test leads for a multimeter are:
   a. usually both black.
   b. usually both red.
   c. colored red for the positive terminal and black for the negative.
   d. colored black for the positive terminal and red for the negative.

8. When using a dc ammeter in a circuit, you should connect the meter so that:
   a. the electron current to be measured enters the negative terminal of the meter and leaves the positive terminal.
   b. the electron current to be measured enters the positive terminal of the meter and leaves the negative terminal.
   c. the positive and negative terminals of the meter are connected directly to the positive and negative terminals of the voltage source.

The Basic Telephone System

ABOUT THIS CHAPTER

You will find that a telephone system is a simple electrical circuit that operates according to principles you have learned earlier. Parts of the telephone circuit convert sound into electrical signals. Other parts change the electrical signals back into sound. As a result, conversations can be transmitted through wires for extremely long distances.

You are already familiar with the mouthpieces and earpieces of a telephone. When you complete your study of this chapter, you will understand how these parts work and how they are connected in an operating telephone system.

THE BASIC PRINCIPLES OF SOUND

Have you ever built a mechanical telephone system using a pair of small coffee cans and a length of string? If you have, you know sound can be transmitted through a string. As crude as this mechanical system is, it demonstrates many of the principles used in modern telephones.

Figure 4-1 shows how a mechanical telephone system operates. Note that the key part of the mechanical telephone system is the flexible metal disk at the bottom of each can. Speaking into the can causes sound waves to strike the disk and make it vibrate. The vibrations from one disk are carried to the other disk by a tightly stretched string. The second disk repeats the in-and-out motions of the sending disk and causes varying pressures in the can. These are sound waves that are crude reproductions of the original sound waves.

Frequency

Frequency is a measure of the number of times per second an object vibrates.

A study of the basic principles of sound reveals how mechanical and electrical telephone systems work. Sound is made up of vibrations. Differences in sound are determined by their "frequency"—the number of times per second a sound vibrates. A high tone (such as a shriek) has a high frequency—several thousand vibrations per second. A low tone (such as a deep bass voice) has a low frequency, on the order of a few hundred vibrations per second.

**Figure 4-1.
A Mechanical
Telephone System**

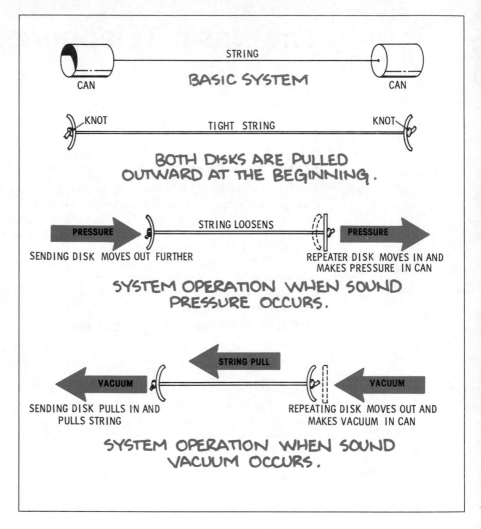

Pitch

Each tone has a specific frequency that is often referred to as its "pitch." A tuning fork, for example, vibrates and creates a sound tone at the frequency for which it was designed. The same is true of pianos and violins, the head of a drum, and your own vocal cords.

Transmission

Air consists of a large number of extremely tiny particles, several million per cubic inch. When sound causes these air particles to vibrate, they alternately pack together and separate at the frequency of the sound. Packing together creates instantaneous areas of high pressure, and the separation develops a condition of less-than-normal air pressure.

Sound travels through air as areas of changing air pressure that can, in turn, cause other objects to vibrate at the same frequency.

As the areas of changing air pressure strike other adjacent air particles, the process is continued. This is the manner in which sound travels through air. When the changes in air pressure strike a flexible disk, or "diaphragm," the disk vibrates. The vibrations are at the same frequency as the original sound.

In a mechanical telephone system, the sending diaphragm transmits its vibrations to a tightly stretched string that, in turn, sets up the same vibrations in the receiving diaphragm. In a modern telephone system, a proper design and use of electricity result in excellent reproduction of sound over long distances.

AN ELECTRICAL TELEPHONE SYSTEM

The basic electrical telephone system consists of a mouthpiece that is connected to a distant earpiece by electrical wires. This system permits conversation in only one direction, however. For two-way conversation, each end of the system requires both a mouthpiece and earpiece.

The mouthpiece in a modern electrical telephone converts sound vibrations into varying electrical current. The earpiece converts varying electrical current into sound vibrations.

The mouthpiece in an electrical telephone system converts sound vibrations into a varying electrical current that can move through a circuit. The earpiece converts those varying electrical signals back into sound vibrations that have the same frequencies as the original sound.

The mouthpiece is sometimes called a "transmitter"; and the earpiece is sometimes called a "receiver." The transmitter sends varying currents through wires to the receiver. As shown in *Figure 4-2*, the receiver converts the varying currents (and voltages) into sounds that we can hear.

The mouthpiece and earpiece in a modern telephone each include a diaphragm that vibrates at sound frequencies.

Although they perform different functions, the mouthpiece and earpiece in an electrical telephone system look very much alike. Both include a diaphragm that vibrates at sound frequencies.

The diaphragm in the mouthpiece, or transmitter, is part of a mechanism that converts sound into fluctuating currents and voltages. The diaphragm in the earpiece, or receiver, is part of a mechanism that converts the fluctuating currents and voltages into sound vibrations of the same frequency.

Figure 4-2.
Sound Conversion in an
Electrical Telephone
System

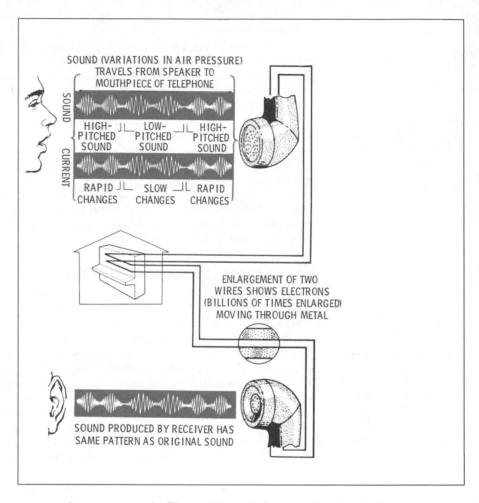

As you can see in *Figure 4-3*, a telephone system is actually a simple electrical circuit.

Figure 4-3.
A Telephone Circuit

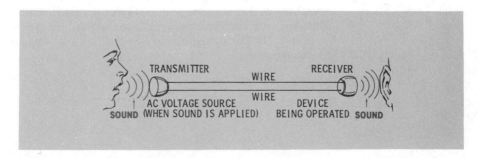

In most commercial telephones, the mouthpiece and earpiece are contained in a single handset. A mechanical view of the main parts of a handset are shown in *Figure 4-4*. *Figure 4-5a* shows the mouthpiece, or transmitter, with no sound applied. *Figure 4-5b* demonstrates what happens when sound strikes the diaphragm.

Figure 4-4.
Parts of a Telephone Handset

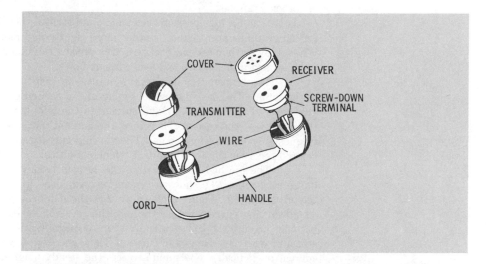

Figure 4-5.
Transmitter Operation

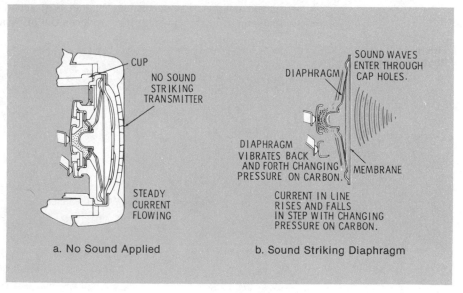

a. No Sound Applied   b. Sound Striking Diaphragm

Your Telephone Company

Your local telephone company provides the electrical power required for operating the system.

When the handset is lifted from the cradle, a steady current from your local telephone company starts to flow through a packet of carbon granules. Sound striking the diaphragm places varying pressure on the carbon. As a result, a varying amount of current flows through the packet of carbon granules.

When a high-pressure area of sound strikes the diaphragm, the carbon granules are packed together more tightly and more current flows. When a lower-than-normal pressure area strikes the diaphragm, the carbon granules are pulled apart and less current flows.

The transmitter diaphragm vibrates in response to the frequency of the sound. The packing and loosening of the carbon granules follow the vibrations of the diaphragm, and current in the telephone line varies with the density of the carbon. Therefore, the current varies at the same frequency as the original sound. The varying current is routed to the desired receiver through a central telephone office.

The method for converting the current back into sound is slightly different. *Figure 4-6a* shows a typical receiver used in a commercial telephone system. *Figure 4-6b* shows that the diaphragm vibrates when a sound-varying current passes through the receiver, or earpiece. The current passing through the coil of wire develops a magnetic field that varies in strength with the changes in current. Thus, the field developed by the current periodically repels and attracts the steady magnetic field of the permanent magnet. The permanent magnet is attached to the diaphragm. The movement of the diaphragm reproduces the original sound.

**Figure 4-6.
Receiver Operation**

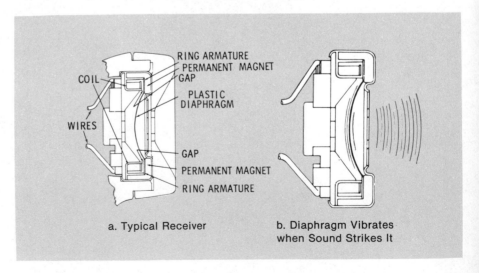

a. Typical Receiver

b. Diaphragm Vibrates when Sound Strikes It

WHAT HAVE WE LEARNED?

1. A basic telephone system uses all the principles of a simple circuit.
2. Sound is matter (such as air) vibrating at a pitch determined by its frequency.
3. In a telephone mouthpiece, or transmitter, sound vibrations in air cause a thin diaphragm to vibrate. The vibrating diaphragm compresses and releases carbon granules. The amount of electrical current through the transmitter varies according to the compression of the carbon granules.
4. In a telephone earpiece, or receiver, changing amounts of current create a changing magnetic field. The magnetic field causes a diaphragm to vibrate, and those vibrations are converted to sound in air.

KEY WORDS

Diaphragm
Frequency
Receiver
Transmitter

Quiz for Chapter 4

1. Which one of the following phrases best describes the nature of sound?
   a. The mechanical vibration of matter (such as air).
   b. Changes in current level in a conductor (such as a wire).
   c. Changes in voltage level at a diaphragm.

2. Which one of the following phrases best describes frequency?
   a. The change in current level through a conductor.
   b. The amount of voltage applied to a diaphragm.
   c. The number of vibrations per second.

3. In a modern telephone system:
   a. the earpiece and mouthpiece are both transmitters.
   b. the earpiece is a transmitter, and the mouthpiece is a receiver.
   c. the earpiece is a receiver, and the mouthpiece is a transmitter.
   d. the earpiece and mouthpiece are both receivers.

4. In a modern telephone system, the transmitter:
   a. converts varying electrical currents into sound vibrations.
   b. converts sound vibrations into varying electrical currents.
   c. uses a diaphragm that vibrates in response to current in a coil of wire.

5. In a modern telephone system, the receiver:
   a. converts varying electrical currents into sound vibrations.
   b. converts sound vibrations into varying electrical currents.
   c. uses a diaphragm that compresses and loosens carbon granules in response to current flowing through a coil of wire.

6. What is the source of electrical power that is used for operating your telephone?
   a. Your local telephone company supplies the power through the wires.
   b. There are batteries built into the housing of your own telephone.
   c. The power comes from the same plug that supplies power for all other appliances in your home.
   d. A modern telephone system does not use electrical power.

7. Which one of the following statements is true?
   a. Compressed carbon granules generate a voltage.
   b. Compressed carbon granules vibrate to produce an audible sound.
   c. Compressed carbon granules can conduct electrical current better than loosened granules.
   d. Loosened carbon granules can conduct electrical current better than compressed granules.

Reading Electrical Diagrams

ABOUT THIS CHAPTER

Before proceeding with a study of more complex circuits, you should learn how to read and draw diagrams used in electricity and electronics. There are many varieties of diagrams, but they have all grown from two basic types: wiring diagrams and schematic diagrams. This chapter explains the meanings of different kinds of diagrams and how you can read and understand them.

ELECTRICAL DIAGRAMS

A technical book can be written without illustrations, but very few are. Words alone cannot fully describe the idea an author wants readers to understand. Writers use diagrams and illustrations along with words to make sure their descriptions are more clearly understood.

Most of the circuits illustrated thus far in this book have been representationl, that is, realistic. A dry cell is drawn as it actually appears —as a cylinder with its terminals in the correct positions. A lamp appears similar to those in your home, and wires are drawn to look as natural as possible.

However, can you imagine trying to understand all the circuits, parts, wires, and terminals in a television set if they are shown in a realistic, three-dimensional form? Such illustrations would be not only difficult to draw but also awkward to use.

Technical drawings are needed by engineers who design equipment, workers who construct it, and technicians who service it. Drawings are also required by people who study electricity and electronics.

Two-Dimensional Diagrams

Electrical diagrams are almost always drawn in a flat, two-dimensional form because they are thus easier to draw and read. Reading a diagram means obtaining important information from it, such as following the path of current through a circuit. Reading a two-dimensional diagram is simplified by eliminating unnecessary and confusing details. But reading this type of diagram is easy only if you understand the language.

The Language of Symbols

Electrical and electronics diagrams use symbols to represent the real item. There are simplified symbols for every electrical and electronic part. When new parts are invented, a corresponding symbol is developed.

Using symbols instead of cumbersome, realistic pictures is not a new idea. Shortly after the Stone Age, people found it difficult to work with their counting and numbering system. Making marks on the ground or stacking pebbles in a pile became fairly tedious when they wished to indicate "how many" of anything. Numeral symbols were invented to show "how many" in a convenient form. This permitted the ancient Arab, who perhaps owned nine sheep and four horses, to show the symbols 9 and 4 on inventory records instead of marks or pictures of the animals themselves.

Learning to interpret electrical and electronic symbols requires the same process you used for learning the meaning of numerals. Learn what the symbols stand for, and you can learn to read any kind of circuit diagram.

WIRING DIAGRAMS

Wiring diagrams are used as guides for constructing circuits and equipment. They are also useful for locating wires and connections when you service a piece of electrical or electronic equipment. A fundamental wiring diagram shows a symbol for each part. Emphasis is placed on showing the terminals of each part and the wire connections between them. A circuit can be easily assembled by following a wiring diagram. Compare the two drawings in *Figure 5-1*.

Electrical and electronics diagrams use simplified symbols to represent the major components in a circuit and the electrical connections between them.

Wiring diagrams emphasize the way wires are connected between components in a circuit.

**Figure 5-1.
Diagrams of a Lamp Circuit**

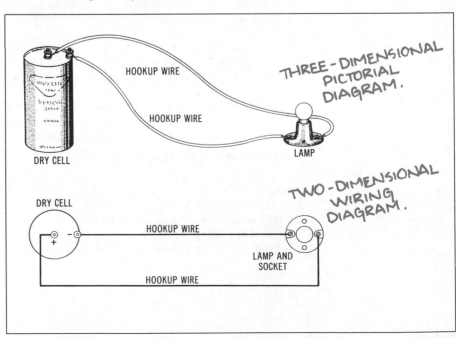

Note how easy it would be to follow the two-dimensional diagram if you were constructing the circuit. The symbols are easily identifiable. The dry cell is a flat circle (top view) with terminals shown in the correct positions and the polarity markings clearly indicated. The lamp symbol shows its two terminals. Wires appear as straight lines. All parts are clearly labeled.

No two manufacturers will necessarily use identical symbols for the same part on a wiring diagram. Each symbol, however, will be a close representation of the real thing. The symbol for the knife switch in *Figure 5-2*, for instance, shows the terminals and clarifies the difference between the open and hinged ends.

Figure 5-2.
Wiring Diagram of a
Lamp Circuit That Uses
a Switch

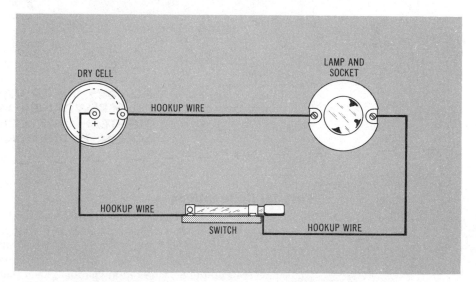

Figure 5-3 is a wiring diagram for two 1.5-V lamps connected to a 1.5-V dry cell. You may construct the circuit if you wish. The circuit is intended to demonstrate a principle of electricity. Note that the symbol for the lamps is slightly different from the one used in the previous wiring diagram. The terminals are shown in the same way, however, and the meaning is equally clear in both cases.

A new part has been added—a terminal board labeled TB1. As shown in its construction detail, a terminal board has a metal strip with one screw at each end. The board itself is made of a tough, plastic insulating material. These boards are convenient connecting points for wires.

Terminal boards are convenient connection points for wires.

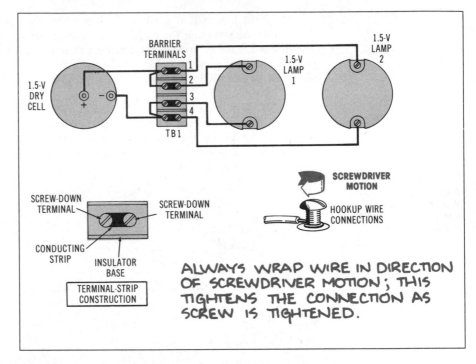

Parallel Connections

The electrical principle demonstrated by the circuit is that two lamps can be connected to a single voltage source. Both lamps are connected across the voltage source; that is, the top terminal of each lamp is wired to the positive pole, and the bottom terminal, to the negative pole. The lamps are said to be connected in "parallel," and that means 1.5 V is applied to each lamp.

The wiring diagram in *Figure 5-4* shows three lamps connected in parallel. Notice that the top terminal on each lamp is connected to the positive terminal on the dry cell. Likewise, the bottom terminal on each lamp is wired to the negative terminal.

This circuit uses three different terminal boards TB1, TB2, and TB3. Each terminal board has two connectors, labeled 1 and 2. Thus any one of the terminal-board connections can be clearly identified by specifying its terminal-board number (TB1, TB2, or TB3) and its individual terminal number (1 or 2).

The top connection on lamp 2 is thus wired to terminal 1 of TB2. By the same token, the positive terminal of the dry cell is wired to terminal 1 of TB1, terminal 1 of TB2, and terminal 2 of TB3.

With just three lamps in the circuit, the diagram is becoming cluttered with lines representing the wires. However, another type of wiring diagram removes much of that clutter.

Figure 5-4.
Conventional Wiring
Diagram

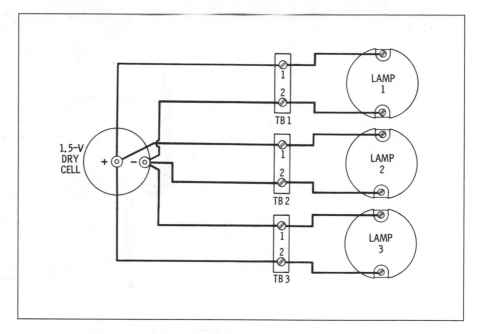

Highway Wiring Diagram

The wiring in *Figure 5-5* looks like a highway with secondary roads leading to and from it. You might call it a highway wiring diagram. Each wire entering the highway has its destination clearly marked. Each wire leaving the highway has labels indicating its source.

The label TB1-1 indicates a wire going to connector 1 of terminal board TB1; TB2-2 specifies a wire going to connector 2 on terminal board TB2.

Although there are only two lines shown at the terminals of the dry cell, the labels assigned to those lines indicate there are three wires in each "bundle."

Figure 5-5.
Highway Wiring
Diagram

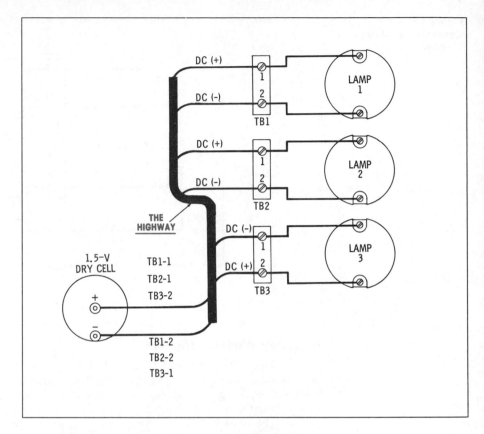

Airline Wiring Diagram

Figure 5-6 shows yet another kind of wiring diagram for the same circuit. This airline version shows wire destinations and sources without a connection between the terminals. An airline diagram is read in much the same way as the highway version. The only difference is that the highway is omitted from the airline drawing. Once you realize that wires actually run between the points indicated by the labels, this version is just as clear as the others.

SCHEMATIC DIAGRAMS

Schematic diagrams are used more than any other technical diagram in electronics. Engineers use schematics (the term "diagram" is usually dropped) when designing equipment and testing its performance after construction. Technicians constantly refer to schematics while servicing or troubleshooting equipment.

Information, including a schematic, is available for nearly every television set, radio, stereo, personal computer, or other electronic device ever manufactured. Most schematics can be purchased at electronics supply stores and from mail order companies.

Figure 5-6.
Airline Wiring Diagram

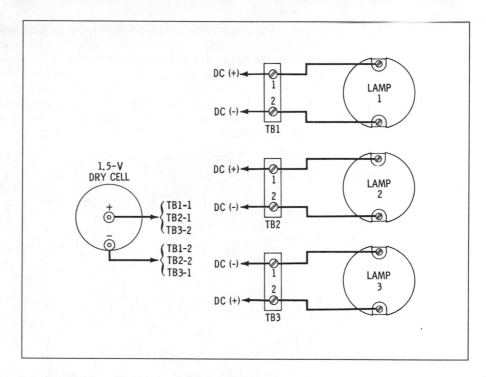

A schematic diagram shows all components and wiring for a circuit. Most components have a standardized symbol.

Schematics are used in nearly all textbooks on electricity and electronics so that all future technicians and engineers will become familiar with the type of diagrams they will be using most often in their work. Another reason is the clarity with which schematics provide information. The many parts of a circuit, or group of circuits, can be drawn in a limited amount of space. Schematic symbols are fairly standard and do not vary as much as the symbols for wiring diagrams.

Figure 5-7 shows the schematic symbol for an incandescent lamp.

Figure 5-7.
Symbols for an
Incandescent Lamp

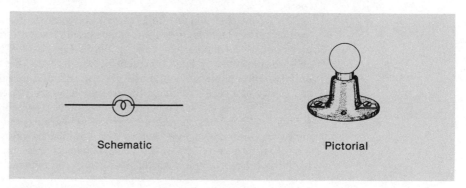

The schematic symbol for a cell is two parallel lines, one being shorter than the other. The shorter line represents the negative (−) terminal, and the longer, the positive (+) terminal. As illustrated in *Figure 5-8*, it is also customary to label the shorter, negative terminal with a minus sign, and the longer, positive terminal with a plus sign. Because a battery consists of two or more cells, its symbol shows a corresponding number of cell symbols. The voltage rating of the battery is shown beside its schematic symbol.

Figure 5-8.
Schematic Symbols for Cells and Batteries

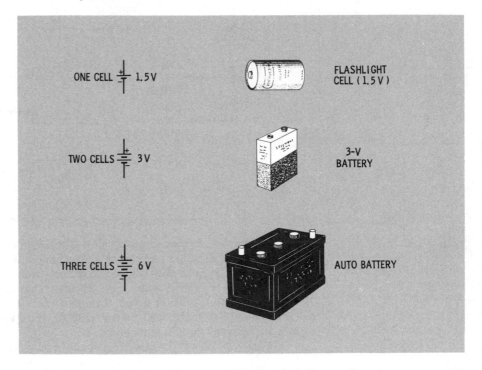

Now that you are familiar with the symbols for a voltage source (cell or battery) and an operating device (incandescent lamp), you should be able to read the simple schematic shown in *Figure 5-9*. The lamp and 1.5-V cell are connected by lines representing wires. Note that the lines run in only two directions: horizontal and vertical. Slanted or curved lines lessen the clarity of a schematic diagram, so they are rarely used.

The schematic symbol for a simple on/off switch, such as the knife switch in *Figure 5-10*, suggests its actual appearance and operation. The arrow on the switch symbol has no real significance except to help to identify it as a switch when shown in the closed position. Otherwise it would look like a simple piece of wire connected between two terminals.

**Figure 5-9.
Simple Circuit
Schematic**

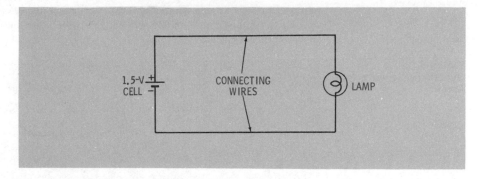

**Figure 5-10.
Schematic Symbols for
an On/Off Switch**

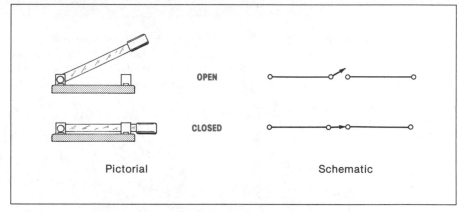

Figure 5-11 is a schematic diagram of a lamp circuit that includes a
switch for turning the light off and on. When the switch is in the open
position (as shown in the diagram), current cannot flow through the circuit,
and the lamp is turned off. Closing the switch completes the circuit,
however, and allows current to flow through the lamp.

**Figure 5-11.
Schematic Diagram of a
Lamp Circuit**

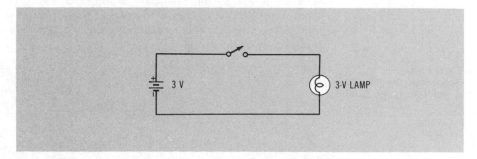

Meters are quite often inserted into circuits to monitor voltage and
current levels. It is thus necessary to show on a schematic where a meter
is to be connected and the type of meter to be used. The symbols for two

types of meters are shown in *Figure 5-12*. Sometimes it is necessary to specify the meter range on the schematic symbols for meters that are inserted into a circuit.

**Figure 5-12.
Schematic Symbols for
Meters**

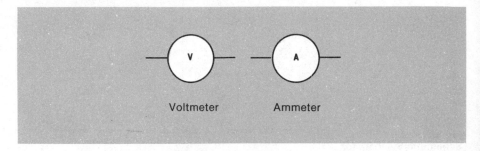

The cell and battery symbols described earlier represent dc voltage and current sources. There are also other types of dc sources—dc generators, for instance. The symbols for a dc generator as well as an ac source of voltage and current are shown in *Figure 5-13*.

**Figure 5-13.
Schematic Symbols for
Voltage Sources**

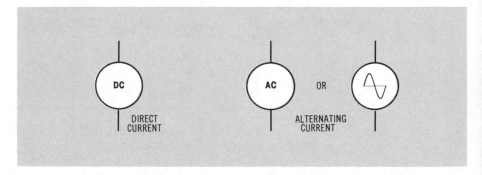

One of the most commonly used sources of ac voltage and current is a simple wall socket. The 120 V ac from that source is often connected to an electrical circuit by means of a plug. *Figure 5-14* shows a 120-V lamp and switch circuit that is connected to a wall outlet by means of an ordinary two-prong plug.

**Figure 5-14.
Schematic Diagram for
a 120-V Ac Lamp Circuit**

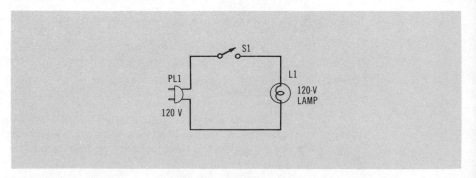

A coil of wire, applied in one way or another to generate magnetic fields, uses a schematic symbol that looks very much like the turns of wire it represents. See the examples in *Figure 5-15*.

**Figure 5-15.
Schematic Symbols for
a Coil of Wire**

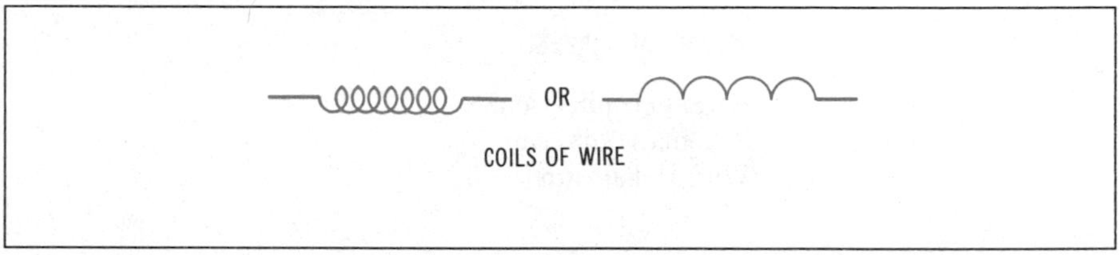

COILS OF WIRE

Quite often one wire is connected to another. On a wiring diagram, the terminal where the connection is made is clearly shown. On a schematic diagram, however, terminals are usually not indicated in a pictorial way. Also, it is sometimes necessary to show lines crossing one another. The illustrations in *Figure 5-16* show how connections and crossings are indicated.

**Figure 5-16.
Schematic Symbols for
Connecting and
Crossing Wires**

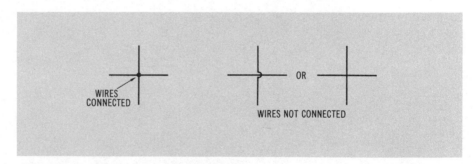

WIRES CONNECTED

OR

WIRES NOT CONNECTED

WHAT HAVE WE LEARNED?

1. Technical diagrams are drawn with symbols to clearly present a great deal of information in a limited amount of space.
2. There are two basic types of diagrams generally used in electrical and electronic work: wiring diagrams and schematic diagrams.
3. Wiring diagrams are useful as a guide when detail is required for construction purposes.
4. Schematic diagrams are widely used by engineers for designing and testing equipment and by technicians for servicing and troubleshooting equipment.
5. A conventional wiring diagram shows each wire, a representative symbol for each part, and a label for each terminal.
6. Schematic symbols have been fairly well standardized.

7. Parts are said to be connected in series when they are connected together in a line.
8. Parts are said to be connected in parallel when they are connected to the same points in a circuit.

KEY WORDS

Parallel connections
Schematic diagram
Wiring diagram

Quiz for Chapter 5

1. A diagram showing how the wiring in your home runs from the fuse box to all of the outlets is most likely:
   a. a wiring diagram.
   b. a schematic diagram.
   c. an anatomical diagram.
   d. an electrical diagram.

2. A diagram showing all the components and wiring in a portable radio is most likely:
   a. a wiring diagram.
   b. a schematic diagram.
   c. a parts list.
   d. a mechanical diagram.

3. Which one of the symbols in *Figure 5-17* represents a coil of wire?
   a. A.
   b. B.
   c. C.
   d. D.
   e. E.
   f. F.

4. Which one of the symbols in *Figure 5-17* represents a switch?
   a. A.
   b. B.
   c. C.
   d. D.
   e. E.
   f. F.

**Figure 5-17.
Figure for Quiz
Questions 3–5**

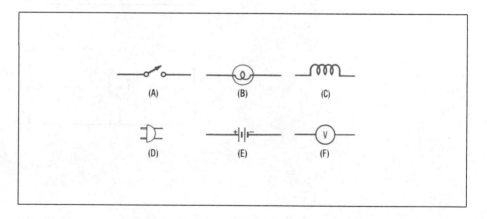

5. Which one of the symbols in *Figure 5-17* represents a wall plug?
   a. A.
   b. B.
   c. C.
   d. D.
   e. E.
   f. F.

6. Referring to circuit A in *Figure 5-18*, suppose that you have inserted the plug into a 120-V

outlet and have set the switch to its off position. What is the status of the lamp and motor?
   a. The lamp and motor are both off.
   b. The lamp is on, but the motor is off.
   c. The lamp is off, but the motor is on.
   d. The lamp and motor are both on.

Figure 5-18.
Figure for Quiz
Questions 6–10

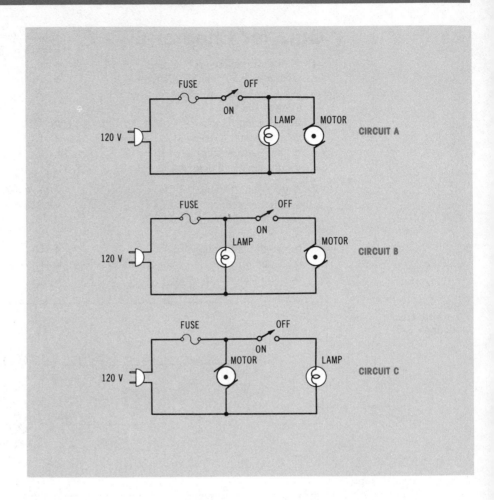

7. Referring to circuit A in *Figure 5-18*, suppose that you have inserted the plug into a 120-V outlet and have set the switch to its on position. What is the status of the lamp and motor?
   a. The lamp and motor are both off.
   b. The lamp is on, but the motor is off.
   c. The lamp is off, but the motor is on.
   d. The lamp and motor are both on.

8. Referring to circuit B in *Figure 5-18*, suppose that you have inserted the plug into a 120-V outlet and have set the switch to its off position. What is the status of the lamp and motor?
   a. The lamp and motor are both off.
   b. The lamp is on, but the motor is off.
   c. The lamp is off, but the motor is on.
   d. The lamp and motor are both on.

9. Referring to circuit B in *Figure 5-18*, suppose that you have inserted the plug into a 120-V outlet and have set the switch to its on position. What is the status of the lamp and motor?

a. The lamp and motor are both off.

b. The lamp is on, but the motor is off.

c. The lamp is off, but the motor is on.

d. The lamp and motor are both on.

10. Referring to circuit C in *Figure 5-18*, suppose that you have inserted the plug into a 120-V outlet and have set the switch to its off position. What is the status of the lamp and motor?

a. The lamp and motor are both off.

b. The lamp is on, but the motor is off.

c. The lamp is off, but the motor is on.

d. The lamp and motor are both on.

Understanding Resistors

ABOUT THIS CHAPTER

Voltage, current, and resistance are closely related within a circuit. Where you find current, you find the other two. Current cannot flow unless there is a force of voltage. The amount of current that flows through a circuit depends on the amount of applied voltage and the amount of resistance to current flow that exists in the circuit. This chapter describes the basic nature of resistance, resistance devices called resistors, and procedures for measuring resistance.

LIMITING THE AMOUNT OF CURRENT FLOW

You have learned that voltage is a force, a kind of electrical pressure, that causes current to flow through a circuit. Generally speaking, the larger the amount of voltage, the larger the amount of current flowing through the circuit.

But consider the fact that all common household light bulbs operate at the same voltage level—about 120 V. Some bulbs are designed to produce more light than others. The brighter the bulb, the more current flowing through it. How is it possible to produce light bulbs that carry more current than others when they all operate from the same voltage level? The reason is that brighter, higher-current light bulbs offer less resistance to current flow than the less bright, lower-current bulbs. "Resistance" is the quality of an electrical or electronic device that limits or controls the flow of current. Resistance (to current flow) is as important to the operation of a device as voltage and current.

RESISTANCE

Resistance is an opposition to current flow in a circuit.

All substances are made up of invisible particles called atoms. There are over 120 different kinds of atoms, or atomic elements. Hydrogen, oxygen, lead, iron, carbon, and uranium are all examples of atomic elements.

One feature that makes one atomic element different from others is the number of electrons it contains. An atom of hydrogen gas, for example, has just one electron. An atom of uranium, on the other hand, has 92.

You know that current is a flow of electrons and that electrons are made to move by the force of voltage. Current is a measure of the rate of flow of electrons under the influence of an applied voltage. This is not to say that an electron leaves the negative pole of a voltage source and moves directly through a circuit to the opposite pole at the speed of light. The motion of electrons through a wire can be compared to pushing a billiard ball into one end of a pipe that is already filled with other billiard balls. The one pushed into one end of the pipe does not immediately fall out of the other end. Rather it causes the ball already located at the opposite end to fall out. Electrons move through a conductor in much the same way.

Even the best conductors offer some resistance to current flow.

Just as there is no such thing as a frictionless bearing for wheels, there is no such thing as a perfect, frictionless conductor. Even the best conductors, such as those having silver or copper atoms, resist the flow of current to some extent. By the same token, there is no such thing as a perfect insulator. If you apply a sufficiently high amount of voltage to an insulator, electrons will begin to flow through it. That is a case of insulation breakdown—usually an undesirable situation.

Some materials conduct electricity better than others. The conductivity of the material depends on how much force is required to make its electrons move. The resistance of a material—its ability to resist current flow—is determined by its atomic structure.

The size of the columns in *Figure 6-1* shows the comparative resistance of certain materials. Keep in mind that no material is a perfect conductor nor a perfect insulator.

Figure 6-1.
Resistance of Materials

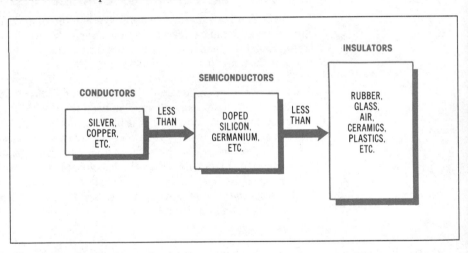

Semiconductors

Semiconductors are materials that are neither good conductors nor good insulators.

Most metals contain atoms that release electrons easily. These materials, therefore, offer the least resistance to current flow. Insulators have the greatest resistance to current flow because their atoms resist the release of electrons. The in-between materials are neither good conductors nor good insulators. Among these are certain materials, called "semiconductors," from which transistors and integrated circuits are manufactured.

Resistance to current flow is measured in ohms.

Resistance to current flow is measured in ohms (abbreviated Ω, the Greek letter omega). The resistance of a typical 1.5-V incandescent lamp, for example, is on the order of 6 Ω. In other words, the lamp offers 6 Ω of resistance to current flow. As a result, applying 1.5 V of electrical pressure causes just 0.25 A to flow through the lamp. If the resistance were lower, more current would flow, and the bulb would burn brighter. On the other hand, if the resistance were higher, less current would flow, and the bulb would not burn as brightly.

Ohm's Law

Because resistance limits the amount of current that flows through a device in response to an applied voltage, there must be some numerical relationship between the voltage, current, and resistance.

Careful experiments would show you that doubling the resistance in a circuit would cut the current in half if the voltage remained unchanged. Current is divided by two when resistance is multiplied by two—assuming, again, that the applied voltage remains the same. Mathematicians would say that current is inversely proportional to resistance. In other words, current decreases as resistance increases.

You can also discover by experimentation what happens to current when voltage increases. You will find that they increase together; that is, current is directly proportional to the amount of voltage. This makes sense because the pressure of voltage causes current to flow. Increasing the amount of pressure increases the flow.

These relationships between voltage, current, and resistance can be expressed in a mathematical statement such as:

Current equals voltage divided by resistance

Using mathematical symbols, this statement becomes:

$$I = \frac{E}{R}$$

where,

I is the current in amperes,
E is the voltage in volts,
R is the resistance in ohms.

If voltage (E) increases, current (I) increases by a proportional amount. By the same token, when voltage decreases, current decreases by a proportional amount. The equation also shows the relationship between current and resistance. Increasing the resistance of a circuit decreases the current flow, and decreasing the resistance increases current flow.

The precise expression of the relationships between current, voltage, and resistance in a circuit is known as Ohm's law. A resistor is a device that provides a specified amount of resistance.

This mathematical expression of the relationship between current, voltage, and resistance is one of the most important in electricity and electronics. It is known as "Ohm's law".

Resistors

Now that you know what resistance is, you are ready to learn about a device called a "resistor." A resistor not only offers resistance to current flow but also has a specific value of resistance. A resistor has many uses, but its main purpose is to control current flow.

Because you know that all materials have some resistance, you should be able to make a resistor having a desired value. The wire from a heating element is an economical material to use. The element with its wire can be purchased from most hardware stores. The wire, called nichrome because it is a mixture of nickel and chromium metals, is also available from appliance repair shops. You can make straight nichrome wire into a coil by winding it tightly around a pencil.

If you make the resistance board shown in *Figure 6-2*, stretch the coil of wire until it is about a foot long. This resistance board can be used to control the current in the familiar lamp circuit. *Figure 6-3* shows both a wiring diagram and a schematic diagram for the circuit. The brightness of the lamp depends on the amount of current flowing through it. The larger the amount of current, the brighter the lamp.

Figure 6-2.
Making a Simple
Resistor

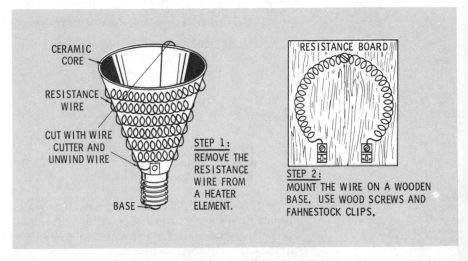

The resistor board can be used for controlling the amount of current that flows through the lamp and, thus, the brightness of the light. If you connect wire 2 directly to terminal D in the circuit, the lamp produces its maximum amount of light. There is no resistance in the circuit to limit the current flow.

Figure 6-3.
A Lamp Control Circuit

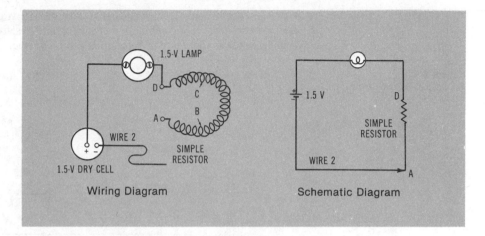

Wiring Diagram

Schematic Diagram

If you connect wire 2 to terminal A, however, the lamp will burn very dimly, if at all. This situation places the maximum amount of resistance in the circuit, thus adjusting the current flow to its lowest level.

Touching the loose end of wire 2 to other places on the resistance wire allows an amount of current to flow that is somewhere between maximum current (no resistance) and minimum current (current flows through the entire length of the resistance). Because this procedure allows you to adjust the amount of current flowing in the circuit, you are also able to adjust the brightness of the lamp.

A potentiometer is a variable resistor.

The arrangement shown in *Figure 6-4* works on the same principle but in a more convenient fashion. A device that can be adjusted to provide a desired amount of resistance is called a "potentiometer." It has a moving contact that performs the function of moving wire 2 across the resistance from terminal A to terminal D. By inserting a potentiometer into a circuit with one wire connected to the movable arm and the other to one of the terminals, a desired value of resistance can be selected.

Figure 6-4.
Making a Potentiometer

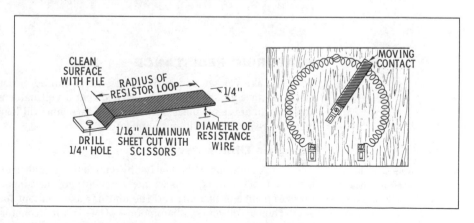

Figure 6-5 shows that the schematic symbol for a potentiometer, or "variable resistor," is a combination of switch and resistor symbols. The arrow indicates that the value of resistance can be varied.

Figure 6-5.
A Lamp Control Circuit

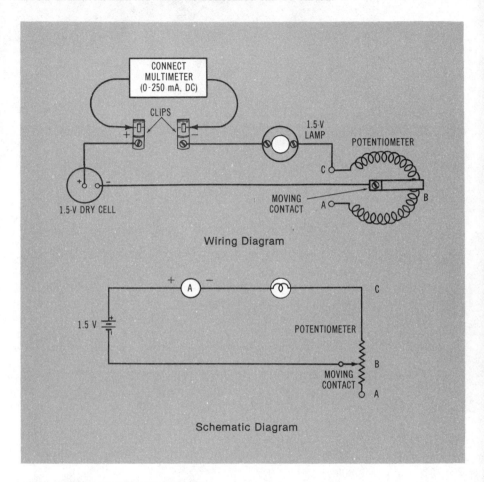

Wiring Diagram

Schematic Diagram

MEASURING RESISTANCE

Like voltage and current, resistance can be measured with a meter. In fact you have already learned that a voltmeter measures voltage and that an ammeter measures current. Here you will learn how to measure resistance with an "ohmmeter."

Using an Ohmmeter

An ohmmeter is an instrument used for measuring resistance.

An ohmmeter is nearly always part of a multimeter. Recall that the volt is the unit of measurement for voltage; the ampere is the unit of measurement for current; and the unit of measurement for resistance is the ohm. Thus a volt-ohm-milliammeter (VOM) is a multimeter that can measure voltage, resistance, and current on the milliampere level.

An ohmmeter scale on a multimeter is labeled in ohms or with the Greek letter omega (Ω) which is often used instead of the word "ohm" after a numerical value of resistance. *Figure 6-6* shows a typical ohms scale on a VOM. The ohmmeter scale reads from infinity (∞) down to 0. Infinity is such a large value that it is impossible to assign a number to it. The K on the scale stands for 1000; therefore 5 K equals 5000 ohms.

Figure 6-6.
An Ohmmeter Scale

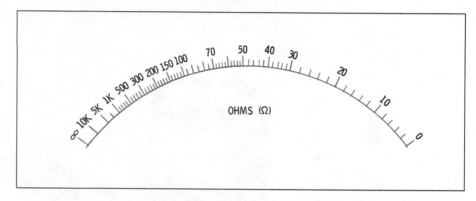

Zero is located on the right side of the scale instead of on the left, as is the case with voltmeters and ammeters. Recall that a meter pointer moves across the dial a distance that is proportional to the amount of current flowing through the meter. If there is no current when a voltage or current measurement is taken, the pointer remains on the left at 0. Maximum voltage or current readings cause the pointer to rest on the right end of their respective scales.

An ohmmeter determines resistance by passing a small amount of current through the resistance being measured.

An ohmmeter measures the value of a resistance by passing current through the resistance. The pointer on the meter responds to the amount of current—the larger the amount of current, the farther the pointer moves to the right.

If the resistance being measured is very small, a relatively large amount of current flows through the resistance and the meter. Thus the pointer moves to the right—in the direction of 0, the smallest possible resistance. On the other hand, using an ohmmeter to measure large amounts of resistance allows smaller amounts of current to flow. As a result the pointer moves a relatively short distance from the highest-possible resistance, infinity.

Bear in mind, however, that because there are no resistances that are perfect conductors nor perfect insulators, there are no resistances that will show exactly zero ohms nor infinite ohms.

An ohmmeter scale is read in much the same way as voltmeter and ammeter scales. If the pointer stops on a numbered division, that number represents the value in ohms. Between numbers, you determine the value of each division mark. Be sure that you reckon the value such that it increases from right to left.

An ohmmeter must be calibrated, or "zeroed," before you can make accurate resistance readings with it. The zeroing procedure, as applied to a multimeter such as the one in *Figure 6-7*, is as follows:

1. Set the function selector to R X 1.
2. Touch the meter probes together. The pointer will rest close to 0 but not always exactly on that value.
3. Slowly adjust the "ohms adj" control knob until the pointer reads exactly 0.

**Figure 6-7.
Calibrating an
Ohmmeter**

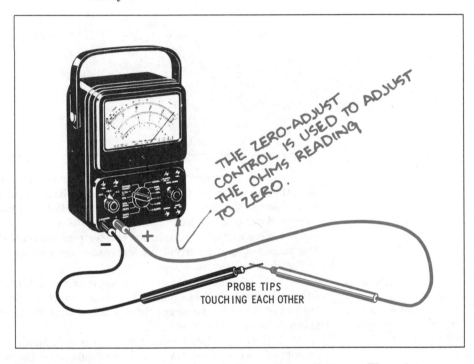

THE ZERO-ADJUST CONTROL IS USED TO ADJUST THE OHMS READING TO ZERO.

PROBE TIPS
TOUCHING EACH OTHER

The ohmmeter is now zeroed and ready for reading resistances. The pointer should swing back to infinity when the probes are not touching each other.

Do not attempt to use an ohmmeter in a circuit that is operating.

Resistance measurements are made by touching the meter probes to the terminals of the unit to be measured. Never make resistance measurements when the circuit under test is operating. Voltage and current readings must be made while the circuit is operating, but the opposite is true for resistance readings. Attempting to use an ohmmeter to measure resistances in an operating circuit can damage the meter.

The schematic diagram in *Figure 6-8* shows how an ohmmeter is used for measuring the resistance of a potentiometer. If the potentiometer contact is at terminal 1 and the meter has been properly zeroed, the resistance reading should be 0 Ω. If you leave the probes connected, the meter will show the gradual increase of resistance as the contact is moved toward terminal 2. At terminal 2, the total resistance of the potentiometer appears on the meter scale.

**Figure 6-8.
Measuring the
Resistance of a
Potentiometer**

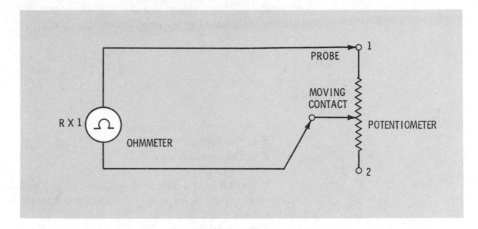

If this experiment is conducted with one of the most popular kind of potentiometers, one-quarter of the total resistance will be read when the contact travels one-quarter of its total rotation, one-half at the halfway point, and so on. This shows that resistance is proportional to the length of the material. If a 12-inch length of wire measures 12 Ω, a 1-inch piece of the same wire measures 1 Ω.

Determining Cold and Hot Resistance

In some cases a resistance taken with an ohmmeter will not be the true resistance of the device when it is operating in a circuit. An incandescent lamp is a good example. It gives off light because current has raised the temperature of the filament to a white-hot level. The physical structure of most materials is such that their resistance increases with a rise in temperature.

Copper, for example, shows an increase in resistance that is proportional to a change in its temperature. Copper is thus said to have a positive "temperature coefficient." Most conductors have a positive temperature coefficient.

Most conductors have a
positive temperature coefficient—their resistance
increases with increasing
temperature.

"Cold resistance" is the ohmmeter measurement taken when operating current is not passing through the device. The heating effect of current is not present. Cold resistance cannot be used to determine current that will be drawn by a heat-generating device.

"Hot resistance" is the true operating resistance of a heated wire or other conductor. It is this resistance that actually determines the amount of current flow. Unfortunately, you cannot measure the hot, operating resistance of such a device because an ohmmeter must not be used in an operating circuit. Recall that current from an operating circuit can flow through an ohmmeter circuit and destroy it.

The hot resistance of a
material must be deter-
mined indirectly, with the
help of Ohm's law.

So how is it possible to determine the operating resistance of an electrically heated device? It can be done indirectly, with the help of an ammeter and a second version of the equation for Ohm's law.

Recall that you can use the following form of Ohm's law to determine the amount of current flowing in a circuit:

$$I = \frac{E}{R}$$

where,

I is the current in amperes,
E is the voltage in volts,
R is the resistance in ohms.

If you know the values of voltage (E) and resistance (R), you can calculate the current (I) by simply dividing the amount of voltage by the amount of resistance.

The second form of Ohm's law looks like this:

$$R = \frac{E}{I}$$

So if you know the values of voltage and current in an operating circuit, you can calculate the operating resistance by dividing the voltage level by the amount of current.

This method of determining resistance is called the voltmeter-ammeter method. *Figure 6-9* illustrates the procedure as applied to a simple incandescent lamp circuit.

**Figure 6-9.
Determining the Hot
Resistance of a Device**

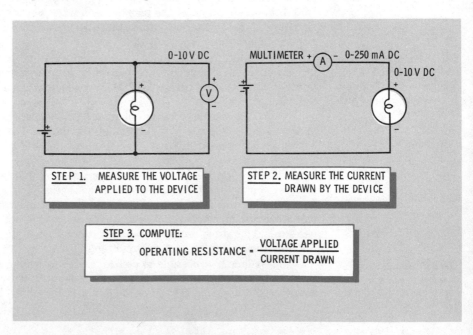

STEP 1. MEASURE THE VOLTAGE APPLIED TO THE DEVICE

STEP 2. MEASURE THE CURRENT DRAWN BY THE DEVICE

STEP 3. COMPUTE:

$$\text{OPERATING RESISTANCE} = \frac{\text{VOLTAGE APPLIED}}{\text{CURRENT DRAWN}}$$

The hot resistance of most conductors is greater than the cold resistance.

Suppose the voltmeter measurement shows the cell applying 1.5 V to the lamp. Further suppose that the ammeter measurement shows 200 mA (0.2 A) flowing through the circuit. According to the equation, if you divide the voltage by the current, the result is the resistance. In this case, 1.5 V divided by 0.2 A equal 7.5 Ω. The same lamp, cooled and removed from the circuit, would show a somewhat lower resistance—its cold resistance.

THE CLASSIFICATION OF RESISTORS

Resistors are classified in two ways: in terms of their construction (wirewound, composition, or film) and in terms of their type or function (fixed, adjustable, or variable). Note the examples in *Figure 6-10*.

**Figure 6-10.
Types of Resistors**

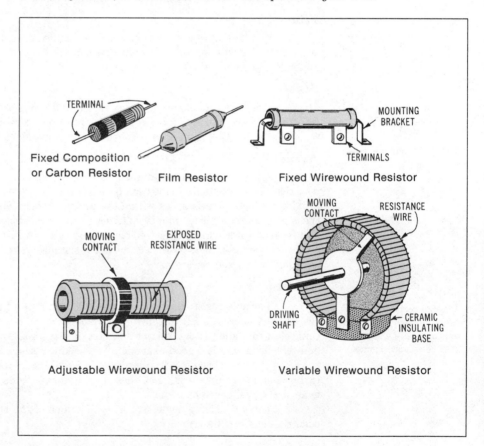

TERMINAL

Fixed Composition
or Carbon Resistor

Film Resistor

MOUNTING
BRACKET

TERMINALS

Fixed Wirewound Resistor

MOVING
CONTACT

EXPOSED
RESISTANCE WIRE

Adjustable Wirewound Resistor

MOVING
CONTACT

RESISTANCE
WIRE

DRIVING
SHAFT

CERAMIC
INSULATING
BASE

Variable Wirewound Resistor

Wirewound

Wirewound resistors are made by wrapping resistance wire around a ceramic or other high-insulation cylinder. The assembly is then covered with enamel glaze and baked. The wire has a known value of resistance per inch.

Composition

Composition, or carbon composition, resistors are molded from a paste consisting of carbon (a conductive material) and a filler material. Terminal wires are inserted into the paste before it hardens. The resistor is then covered with a plastic coating. The resistance and wattage ratings are determined by the ingredients (carbon and filler material) as well as by the diameter and length of the resistor.

Incidentally, carbon is one of the few commonly available conductive materials that has a "negative temperature coefficient." Heating a carbon material causes its resistance to decrease. Recall that most other conductive materials, including resistance wire, have a positive temperature coefficient—their resistance increases with temperature.

Film

Film resistors are made of a layer of resistive material that is applied to an insulating core. The thickness of the layer of resistive material determines the resistance value.

Fixed

Fixed resistors have a specified value of resistance that cannot be changed in any convenient way.

The wirewound, carbon composition, and film resistors just described are examples of fixed resistors. Fixed resistors have a specified resistance value that cannot be changed in any convenient way.

Adjustable

Adjustable resistors, on the other hand, are constructed in such a way that their resistance values can be varied easily. Adjustable resistors provide a range of resistances within the limits of their total value. When these resistors are placed in a circuit, the sliding contact can be positioned and secured to provide the desired resistance value. Adjustable resistors are often used in situations that require a resistance value that is not available with fixed resistors.

Variable

Variable resistors are designed in such a way that you can easily change their resistance values.

Variable resistors are designed for continuous adjustment of their resistance values. A shaft to control the resistance value is usually connected to a knob of one sort or another. The volume control on radios and television sets is a good example of a variable resistor.

The resistance material in variable resistors can be wire or a carbon-composition material. In either case, a moving contact presses against the resistance material.

Table 6-1 shows some typical applications of the three different constructions just described.

**Table 6-1.
Resistor Types and
Applications**

Type	Applications
Composition or Carbon	Composition resistors are the least expensive of the types discussed. They are, therefore, the type most widely used. However, composition resistors have certain limitations. They cannot handle large currents, and their measured values may vary as much as 20% from their rated resistance.
Wirewound	Wirewound resistors are more expensive to manufacture. They are used in circuits which carry large currents or in circuits where accurate resistance values are required. Wirewound resistors can be made to within 99% or better of the desired value.
Fixed Film	Smaller and less expensive than wirewound resistors, film-type resistors have the same accurate resistance values, but not the same large current capability.

THE POWER RATING OF A RESISTOR

As you already know, current flowing through a resistor generates heat. If too much heat is generated, the resistor will be damaged. Wire in wound resistors, for instance, will melt and burn. Some of the carbon in composition resistors will burn away.

The current-carrying capacity of a resistor is rated according to the amount of heat it can safely release in a given period of time. A resistor must not be used in a circuit where current causes heat to build up faster than the resistor can dissipate it. Even if the resistor isn't actually destroyed by the excess heat, the heat may cause a permanent change in resistance value. In addition, heat from an overloaded resistor may damage other components that are nearby.

Power dissipation, usually in the form of heat energy, is measured in watts.

Because heat is a form of energy, the heat-releasing rate of a resistor is measured in terms of its power rating. The unit of measurement for electrical power is the "watt" (abbreviated W). A 100-W light bulb, for example, dissipates 100 W of heat. In the process, the lamp also gives off light.

The power dissipation of a resistance depends on the amount of current flowing through it. The arithmetic of the matter looks like this:

$$P = I \times I \times R$$

where,

P is the power in watts,
I is the current in amperes,
R is the resistance in ohms.

Suppose a 10-Ω resistance has 3 A of current flowing through it. What is the amount of power being dissipated? From the power equation:

$$P = 3 \times 3 \times 10 = 90$$

The amount of power dissipated in this example is 90 W.

Composition resistors are available with power ratings of 0.25 W, 0.5 W, 1 W, and 2 W. You should use wirewound resistors for handling larger amounts of power.

A design engineer determines the values for resistance and current needed in a circuit and then specifies the power rating of the resistor according to the preceding equation. If the value falls between two of the standard power ratings, the engineer always selects the one with the next-higher power rating.

THE TOLERANCE RATING OF A RESISTOR

A resistor rarely measures the exact resistance specified for it. The amount the value varies is called "tolerance." Every resistor has a given tolerance rating as a resistance value and power rating.

The tolerance specification (in percentage) for a resistor is a measure of how much its actual value can vary from its specified value.

Resistor tolerance is specified in terms of a percentage that indicates the amount that the resistance varies above and below its rated value. Standard tolerances for composition resistors are 5%, 10%, and 20%. Wirewound resistors have tolerances as low as 1% and 2%.

Consider a 1000-Ω resistor that has a tolerance of 10%. Ten percent of 1000 is 100. The actual value of the resistor can be as much as 100 Ω above and 100 Ω below its rated value. That means the resistor is considered "good" as long as its value is between 900 and 1100 Ω.

Some kinds of high-precision electronic circuits require resistors having small tolerances. Less critical circuits can operate quite well with resistors that have relatively wide tolerances. You will find that the variety of resistance values available for resistors having lower tolerances is wider than for those having higher tolerance ratings.

When you are purchasing fixed resistors, you should be prepared to describe each of them according to four different characteristics:

1. Type (wirewound, composition, or film)
2. Value (in ohms)
3. Tolerance (1%, 5%, 10%, or 20%)
4. Power rating (0.25 W, 0.5 W, 1 W, or 2 W)

THE SPECIFICATIONS OF A RESISTOR

Wirewound and film resistors normally have their resistance value and tolerance stamped directly on them. Carbon composition resistors, however, use a series of colored bands as labels for their resistance and tolerance ratings.

Color Coding

The resistance value of a composition resistor is usually shown as bands of color painted around the body of the resistor.

Three color bands painted around the body of a resistor specify its resistance. A fourth band (or absence of a fourth band) indicates its tolerance. See the positions of the bands in *Figure 6-11*.

Figure 6-11.
Color Bands for Carbon-Composition Resistors

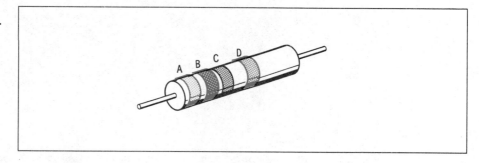

The first three color bands on a resistor indicate its rated resistance value. The fourth color band indicates the tolerance.

The series of colored bands are always located closer to one end of the resistor than to the other. The coding is read, beginning from the band closest to one end of the resistor (band A in the illustration).

Resistance Bands

Regarding the three resistance bands, each represents a value between 0 and 9. The first two bands (A and B in the illustration) represent the first two digits in the resistor value. The third band (C) represents the number of zeroes that follow the first two digits.

Table 6-2 shows the numeric values assigned to the ten different colors.

Table 6-2.
Resistance Color Codes

Color	Number	Color	Number
Black	0	Green	5
Brown	1	Blue	6
Red	2	Violet	7
Orange	3	Gray	8
Yellow	4	White	9

Tolerance Band

The fourth color band represents the tolerance value. Composition resistors that use this color-coding scheme are normally available in tolerances of 5%, 10%, and 20%. So the color-coding scheme for the tolerance band does not have to be as elaborate as that for the resistance bands. A gold fourth band indicates a tolerance of 5%, a silver band indicates a tolerance of 10%, and the absence of a fourth color band indicates a tolerance of 20%. *Table 6-3* summarizes this.

Table 6-3.
Tolerance Color Codes

Color	Tolerance
Gold	5%
Silver	10%
None	20%

Example

Suppose you are looking at a composition resistor that has the following series of color bands: Band A is yellow; Band B is violet; Band C is orange; and Band D is gold.

According to the color-coding scheme, the first two bands indicate the first two digits in the resistance value: yellow is 4 and violet is 7. So the first two digits in the resistance value are 47. The third color band indicates the number of zeroes to follow the first two digits. Orange is 3, so the first two digits should be followed by three zeroes. In other words, the resistor has a value of 47,000 Ω (or 47 K). The fourth color band is gold. According to the table, a gold band indicates a tolerance of 5%. The resistor has a resistance value of 47 K and a tolerance of 5%.

By way of another example, suppose you have a resistor that has just three color bands: red, violet, and black. The first two digits in the resistance value are 27. Because the third color band is black, you must conclude that there are no zeroes following those two digits. It is a 27-Ω resistor. And because there is no fourth color band, you can assume that it has a tolerance of 20%.

SERIES, PARALLEL, AND COMBINATION CIRCUITS

There are only three different ways in which electrical or electronic parts, including resistors, can be connected together in a circuit: in series, in parallel, or in combinations of series and parallel connections. Being able to distinguish series and parallel portions of a schematic diagram can go a long way toward helping you understand how the circuit operates.

Series Circuits

Figure 6-12 shows a series circuit and summarizes its main characteristics. This series circuit is composed of a voltage source and two resistors. The list of definitions shows one of the standardized procedures for describing the voltages, currents, and resistances in the circuit.

**Figure 6-12.
Series Circuit**

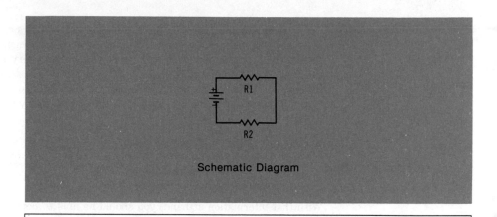

Schematic Diagram

DEFINITIONS

E_T = Total applied voltage (volts)
E_1 = Voltage across resistor R_1 (volts)
E_2 = Voltage across resistor R_2 (volts)
I_T = Total circuit current (amperes)
I_1 = Current through resistor R_1 (amperes)
I_2 = Current through resistor R_2 (amperes)
R_T = Total circuit resistance (ohms)
R_1 = Resistance value of resistor R_1 (ohms)
R_2 = Resistance value of resistor R_2 (ohms)

CHARACTERISTICS

The same current flows through all parts of the circuit: $I_1 = I_2 = I_T$. The total voltage is equal to the sum of the voltages across the individual resistors: $E_T = E_1 + E_2$. The total resistance is equal to the sum of the individual resistances: $R_T = R_1 + R_2$.

In a series circuit, the same current flows through all parts of the circuit, and the source voltage is divided among the individual resistances. The total resistance is equal to the sum of the individual resistances.

The figure describes the essential characteristics of a series circuit—first in words, then in a corresponding mathematical fashion. Take special note of the fact that the same current flows through every part of the circuit and that the total voltage is divided between the two resistors. The total resistance is equal to the sum of the individual resistance values.

Suppose you are working with this circuit and you know:

1. The total current: $I_T = 0.1$ A.
2. The voltage across R_1: $E_1 = 2$ V.
3. The value of resistor R_1: $R_1 = 20\ \Omega$.
4. The value of resistor R_2: $R_2 = 40\ \Omega$.
5. The source voltage: $E_T = 6$ V.

With that information at hand, you should be able to determine:

1. The current through R_1: $I_1 =$ _____ A.
2. The current through R_2: $I_2 =$ _____ A.
3. The voltage across R_2: $E_2 =$ _____ V.
4. The total resistance: $R_T =$ _____ Ω.

You can determine the current through the two resistors by recalling that the same current flows through each part of the circuit. The total current for this example is given as 0.1 A, so you can correctly conclude that 0.1 A flows through the two resistors as well. In other words, $I_1 = 0.1$ A, and $I_2 = 0.1$ A.

Because the total voltage is equal to the sum of the voltages across the individual resistances, you can conclude that the voltage across resistor R_2 is equal to the total voltage minus the known voltage across resistor R_1:

$$E_T = E_1 + E_2$$
$$6\ V = 2\ V + E_2$$
$$E_2 = 6\ V - 2\ V$$
$$E_2 = 4\ V$$

The total resistance of a series circuit is equal to the sum of the individual resistances. In this example the total resistance is $20\ \Omega + 40\ \Omega$, or $60\ \Omega$.

Parallel Circuits

Figure 6-13 shows a parallel circuit and summarizes its main characteristics. This parallel circuit is composed of a voltage source and two resistors. Notice that the definitions are the same as those used for series circuits. However, the list of characteristics is quite different.

**Figure 6-13.
Parallel Circuit**

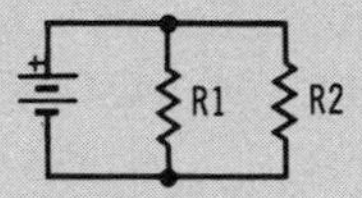

Schematic Diagram

DEFINITIONS

E_T = Total applied voltage (volts)
E_1 = Voltage across resistor R_1 (volts)
E_2 = Voltage across resistor R_2 (volts)
I_T = Total circuit current (amperes)
I_1 = Current through resistor R_1 (amperes)
I_2 = Current through resistor R_2 (amperes)
R_T = Total circuit resistance (ohms)
R_1 = Resistance value of resistor R_1 (ohms)
R_2 = Resistance value of resistor R_2 (ohms)

CHARACTERISTICS

The same voltage appears across all parts of the circuit: $E_1 = E_2 = E_T$. The total current is equal to the sum of the currents through the individual resistors: $I_T = I_1 + I_2$. The total resistance is less than the value of the smaller resistor: $R_T = (R_1 \times R_2)/(R_1 + R_2)$.

Notice that the same voltage appears across every part of the circuit but that the source current is divided among the individual components. The total resistance is less than the value of the smaller resistor. There are a couple of ways to go about calculating the total resistance of a parallel circuit. One way is to use the "product-over-sum" rule:

$$R_T = \frac{(R_1 \times R_2)}{(R_1 + R_2)}$$

This equation literally says: The total resistance of two resistors connected in parallel is equal to the product of the two values divided by the sum of the two values.

The product-over-sum rule for determining total resistance works only for two resistances at a time. When you encounter a circuit that has more than two resistances connected in parallel, you can use the product-over-sum rule for two resistances at a time until the entire circuit is reduced to a single equivalent value—the total resistance value for the circuit.

Suppose you are working with this circuit and you know:

1. The total voltage: $E_T = 6$ V.
2. The current through R_1: $I_1 = 0.3$ A.
3. The value of resistor R_1: $R_1 = 20 \ \Omega$.
4. The value of resistor R_2: $R_2 = 40 \ \Omega$.
5. The source current: $I_T = 0.45$ A.

With that information at hand, you should be able to determine:

1. The voltage across R_1: $E_1 =$ _____ V.
2. The voltage across R_2: $E_2 =$ _____ V.
3. The current through R_2: $I_2 =$ _____ A.
4. The total resistance: $R_T =$ _____ Ω.

You can determine the voltage across the two resistors by recalling that the same voltage appears across each part of a parallel circuit. The total voltage in this example is given as 6 V, so you can conclude that 6 V appears across the two resistors as well: $E_1 = 6$ V, and $E_2 = 6$ V.

Because the total current is equal to the sum of the currents in the individual branches of the circuit, you can determine that the current through resistor R_2 is equal to the total current minus the current through resistor R_1. In this example, $I_2 = 0.45$ A $- 0.3$ A, or 0.15 A.

You can use the product-over-sum rule to calculate the total resistance of this parallel circuit:

$$R_T = (R_1 \times R_2)/(R_1 + R_2)$$
$$R_T = (20 \times 40)/(20 + 40)$$

$$R_T = 800/60$$
$$R_T = 13.3 \; \Omega$$

Combination Circuits

A combination circuit is one that has some parts connected in series, and others in parallel. *Figure 6-14* shows two examples of combination circuits.

**Figure 6-14.
Combination Circuits**

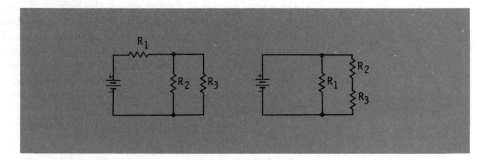

WHAT HAVE WE LEARNED?

1. Resistance is the opposition to current flow in a circuit.
2. Even the best conductors offer some resistance to current flow.
3. Semiconductor materials are neither good conductors nor good insulators.
4. Ohm's law expresses the exact relationship between current (I), voltage (V), and resistance (R):

$$E = \frac{I}{R} \; .$$

5. A resistor is a device that provides a specified amount of resistance.
6. A potentiometer is a variable resistor. It has a shaft that you can rotate in order to set a desired amount of resistance.
7. Resistance can be measured with an ohmmeter.
8. An ohmmeter determines resistance by running a small amount of current through the resistance being measured.
9. An ohmmeter should be calibrated, or "zeroed," before it is used.
10. Attempting to use an ohmmeter to measure resistance in an operating circuit can damage the meter.
11. Most conductors have a positive temperature coefficient. This means resistance increases as the temperature of the conductor increases.
12. A few conductors, including carbon, have a negative temperature coefficient. This means the resistance decreases as the temperature increases.
13. The cold resistance of a conductor is the amount of resistance it has when power is not applied to it. Cold resistance of heating elements can be measured with an ohmmeter.
14. The hot resistance of a heating element is the resistance it has when electrical power is applied to it. If the material has a positive temperature coefficient, the hot resistance is higher than the cold resistance.

17. Electrical power is specified in units called watts.
18. Resistors are manufactured with specific power ratings. The power dissipation of a resistor should not exceed its power rating.
19. The tolerance of a resistor is a measure of how much the actual resistance value can vary from the rated resistance value.
20. Composition resistors use color bands to indicate their resistance value and tolerance.
21. In a series circuit, the same current flows through all parts of the circuit, and the source voltage is divided among the individual resistances. The total resistance is equal to the sum of the individual resistances.
22. In a parallel circuit, the same voltage appears across all parts of the circuit, and the total current voltage is divided among the individual branches. The total resistance is less than the value of the smaller resistor.
23. The product-over-sum rule states that the total resistance of two resistors connected in parallel is equal to the product of the values divided by the sum of the values.
24. A combination circuit is one that has some components connected in series and others in parallel.

KEY WORDS

Cold resistance
Fixed resistor
Hot resistance
Ohm's law
Parallel circuit
Potentiometer
Resistance

Resistor
Semiconductor
Series circuit
Temperature coefficient
Tolerance
Variable resistor

Quiz for Chapter 6

1. Resistance is:
   a. the force that causes current to flow in a circuit.
   b. the rate of electron flow in a circuit.
   c. the opposition to current flow in a circuit.
   d. the rate of dissipation of energy in an electrical device.

2. Resistance is measured in:
   a. volts.
   b. amperes.
   c. ohms.
   d. watts.

3. Which one of the following statements most accurately describes Ohm's law?
   a. Current in a circuit is proportional to both the applied voltage and the amount of resistance.
   b. R = I/E.
   c. Current in a circuit is proprotional to the amount of applied voltage but inversely proportional to the amount of resistance.
   d. The total resistance of a circuit is equal to the sum of the individual resistances.
   e. What goes up must come down.

4. Which kind of resistor is used as a volume control in audio equipment?
   a. Fixed resistor.
   b. Adjustable resistor.
   c. Variable resistor.
   d. Audio resistor.
   e. Video resistor.

5. The instrument used for measuring resistance is called:
   a. an ohmmeter.
   b. a voltmeter.
   c. an ammeter.
   d. a wattmeter.

6. A material that has a positive temperature coefficient:
   a. expands when heated.
   b. contracts when heated.
   c. generates more voltage when heated.
   d. has a higher resistance when heated.
   e. has a lower resistance when heated.

7. How much current flows through a 10-Ω resistance that has 120 V applied to it?
   a. 0.0833 A.
   b. 8.33 A.
   c. 12 A.
   d. 1200 A.

8. What is the hot resistance of a heating element that draws 1.5 A at 120 V?
   a. 0.0125 Ω.
   b. 1.25 Ω.
   c. 80 Ω.
   d. 180 Ω.

9. The wattage rating of a resistor is an indication of:
   a. the amount of power it can safely dissipate.
   b. how much the actual resistance value might vary around the rated value.
   c. the amount of opposition to current flow.
   e. whether it should be used in a series or parallel circuit.

10. A certain 2000-Ω resistor has a tolerance of 5%. The resistor is considered "good" if its actual resistance is between:
   a. 100 and 200 Ω.
   b. 1800 and 2200 Ω.
   c. 1900 and 2100 Ω.
   d. 2000 and 2100 Ω.
   e. 2000 and 2200 Ω.

11. What is the specified resistance value for a resistor that has color bands of red, orange, red, and silver?
   a. 0.0032 Ω.
   b. 232 Ω.
   c. 230 Ω.
   d. 2300 Ω.
   e. 2530 Ω.

12. Which one of the following statements applies to all series circuits?
   a. The same current flows through all components.
   b. The same voltage is applied to all components.
   c. The total resistance is less than the value of the smallest resistor.
   d. The total resistance is greater than the sum of the individual resistances.

13. Which one of the following statements applies to all parallel circuits?
   a. The same current flows through all components.
   b. The same voltage is applied to all components.
   c. The total resistance is equal to the sum of the individual resistances.
   d. The total resistance is greater than the sum of the individual resistances.

14. Which one of the following statements represents the results of applying the product-over-sum rule to resistance values?
   a. The total resistance is equal to the sum of the individual resistances.
   b. The total resistance is greater than the sum of the individual resistances.
   c. The total resistance is less than the sum of the individual resistances but greater than the value of the smallest resistance.
   d. The total resistance is less than the value of the smallest resistance.

How to Solder

ABOUT THIS CHAPTER

This chapter describes how to make permanent electrical connections using a procedure called soldering. You will learn how to work with the necessary tools and hardware. The fundamentals are simple, but you must practice in order to attain the level of skill required for doing the job properly.

When you complete the work suggested in this chapter, you will be able to select the proper soldering tools, prepare a soldering iron, make proper mechanical and electrical connections, make good solder joints, and remove the solder from a connection when necessary.

THE PURPOSE OF SOLDERING

"Solder" is a special mixture applied to electrical connections to prevent oxidation. Soldering is the process of applying the right amount of solder to join two or more pieces of metal.

Oxidation

Metals are good conductors, but their exposure to air creates oxides, which are poor conductors.

Oxidation is the result of a chemical reaction between air and certain metals. When iron oxidizes, for instance, rust forms on its surface; when copper oxidizes, a dull film forms on its surface. Although most metals are good conductors, oxidized metals are usually very poor insulators. In order that the desired conductive nature of metals in electrical circuits be retained, any oxidation must be removed and prevented from forming in the future.

Solder on an electrical connection prevents oxidation of the metals and improves the mechanical strength of the connection.

One way to prevent steel from oxidizing is to paint it. Autos are painted partly for the sake of appearance but mainly to prevent the formation of damaging oxidation in the form of rust. Copper and other metals used for making electrical connections must be protected from oxidation. Electrical connections are not painted, but they are coated with a protective layer of solder. In addition to preventing the formation of oxidation at an electrical connection, solder provides additional mechanical strength at the point where wires join a terminal.

SOLDER

Solder is applied to a connection by melting the solder. Soldering irons and soldering guns are tools used to apply the necessary amount of heat.

Solder is a metal alloy (mixture) containing tin and lead. The alloy most often recommended for soldering in electronic circuits is one that uses a mixture of 60% tin and 40% lead. This is called 60/40 solder. Other tin/lead mixtures are available, many intended for uses other than electronics—making jewelry and soldering together copper pipe, for example.

Generally speaking, the higher the tin content of a solder, the better suited it is for electronics applications. However, raising the tin content also raises the melting temperature, which risks damaging delicate electronic parts as they are soldered to a circuit. Thus, the 60/40 ratio is the best practical balance between the need for a properly soldered joint and the amount of heat that most electronic components can tolerate.

Solder is available in spools of wire-like material. Small-diameter solder is necessary when soldering small components to circuit boards and joining fine wires to plugs. A large-diameter solder is desirable for making larger solder connections.

Use only 60/40 solder for electronics applications. This is a mixture of 60% lead and 40% tin.

ROSIN

Rosin-Core Solder

Solder that is intended for electronics applications has a hollow core that is filled with a pine-tar derivative called "rosin." This product is called rosin-core solder. When you heat this kind of solder to make a solder connection, the rosin melts out of the core and spreads out over the metals. The purpose is to prevent oxidation during the soldering process itself. Without this rosin, the heating effect of your soldering iron or gun would speed up the oxidation process, thereby making it impossible to avoid getting oxides of the metals mixed into the solder connection. Oxides in the connection reduce its ability to conduct electricity. A secondary purpose of the rosin is to wash, or burn, away small amounts of foreign materials that might be on the connection.

Rosin prevents oxidation during the soldering process and removes small amounts of foreign materials from the solder connection.

Rosin for electrical soldering applications is also available separately in paste and liquid forms. These forms are used only in special applications that require more rosin than can be supplied by a rosin-core solder.

Acid-Core Solder

Old-timers call the rosin material soldering flux. This is a general term that applies to a variety of different materials that prevent oxidation during the soldering process. The flux used in plumbing applications is an acid paste. You can buy acid-core solder from hardware stores, but you must never use it for electronics applications. Acid-core solder might give you a clean, firm solder joint, but it also eats away the metals and ruins the conductive properties of the joint.

Never use acid-core solder for electrical applications.

THE PROPER SOLDERING TOOLS

Soldering Irons and Soldering Guns

The main soldering tool is a soldering gun or soldering iron. Either can be used, but each has certain advantages for specific kinds of jobs. Soldering guns, pistol-shaped tools with a trigger-like switch, provide heat within a few seconds at a touch of the trigger. Because they are larger and heavier than most soldering irons, soldering guns are more difficult to use in close quarters.

Soldering irons reach their operating temperatures slowly and must be unplugged to cool. But they are generally less expensive and simpler to use than soldering guns.

The heating power of soldering guns and irons is rated in watts. Irons are available with ratings from 6 W or less to 500 W or more. The larger irons are for heavy-duty industrial applications. Irons rated between 10 and 75 W are best for most electronics applications. Remember to use no more heat than is necessary to get a properly soldered connection.

Figure 7-1 shows that soldering irons are provided with changeable heating elements and tips. This feature offers several important advantages: you can change the heating element to get the proper heating level for the soldering task at hand, and you can change the tip in order to use the shape that is most appropriate. Another advantage is that you can replace burned-out heating elements and tips without having to replace the entire soldering iron.

Soldering irons rated between 10 and 75 W are best for most electronics applications.

**Figure 7-1.
Replaceable Parts of a
Soldering Iron**

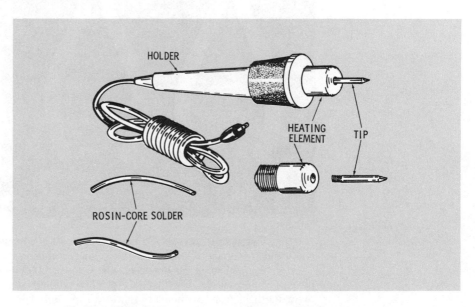

You will often hear that a good, tight mechanical connection is required before soldering. This means you must make a tightly bound connection between the wires and terminal to be soldered. The parts should remain tightly fixed together during the soldering process. If they are properly fixed then, you can be certain they will remain fixed after the job is done.

You must have some tools in addition to a soldering gun or iron to do a proper job. These additional tools include those necessary for cutting, stripping, bending, and crimping the kind of wire you are using. *Figure 7-2* shows three such tools.

Figure 7-2.
Tools for Cutting,
Stripping, Bending, and
Stripping Wire

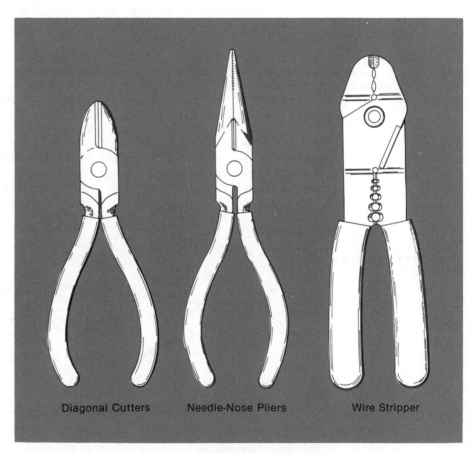

Diagonal Cutters   Needle-Nose Pliers   Wire Stripper

Diagonal cutters, or "dikes," are used for cutting wire, trimming leads and terminals to length, and stripping insulation from wire.

Long-nose, or needle-nose, pliers are used for holding materials in place, forming wire to the shape of terminal connections, splicing wires, and diverting heat from components that are especially sensitive to heat.

Although it is possible to use diagonal pliers for stripping insulation from wire, a special wire-stripping tool can do the job faster and without nicking the conductive part of the wire.

A soldering-iron stand is a convenient accessory that provides a safe and secure holder for a hot soldering iron. Of course you can simply lay a hot soldering iron on the workbench, but you will wish you had used a soldering-iron stand when any one of a wide range of potential accidents occurs—burning your arm, pulling the soldering iron off the bench, or sliding it against a piece of valuable equipment and burning a hole before you are aware of the problem.

A soldering aid, shown in *Figure 7-3*, is a simple and inexpensive tool. It usually has a hook at one end and a sharp point at the other. It can be used for hooking and pulling fine wires, forming them around a terminal, and pushing them into place. You will find that the value of a soldering aid increases with your own soldering expertise.

Figure 7-3.
Soldering Aid

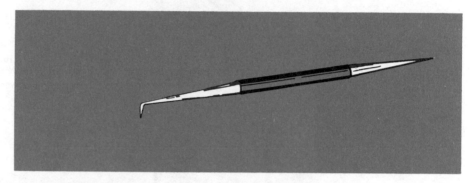

MAKING THE MECHANICAL CONNECTIONS
Preparing the Wire

The mechanical connection, and not the solder, is supposed to hold the wires and terminals together.

Failure to make a clean, tight mechanical connection before is just the first of many careless mistakes that can result in a poor solder connection. The mechanical connection, and not the solder, is supposed to hold the wires in place.

Most wire is supplied with a vinyl insulation that must be removed before making the mechanical connection. You must make sure that you do not nick the conductor when you strip the insulation, and you must make sure that none of the insulation material gets into the solder.

If the stripped portion of the wire is not shiny, it probably has a film of oxidation on it. This film must be removed by carefully scraping the surface with the edge of a sharp knife or razor blade or, better, by sanding it with a very fine grade of sandpaper. Use the same procedure for stripping the insulation from wire that has an enamel, rather than vinyl, insulation.

Never use steel wool to clean electrical connections.

As tempting as it might be, never use steel wool for scraping or cleaning operations. Steel wool leaves countless bits of steel all over the place and risks forming a short circuit between closely spaced conductors on circuit boards and in delicate plugs and connectors.

The amount of insulation you remove depends on the physical size of the connector and wire. You will have a better idea about how much to remove when you see how the stripped area is treated in *Figure 7-4.*

Figure 7-4.
Connecting a Wire to a
Terminal

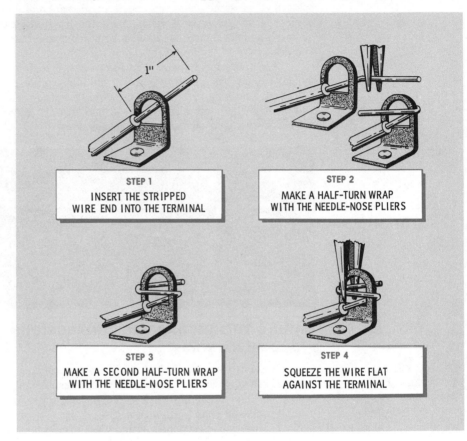

STEP 1
INSERT THE STRIPPED
WIRE END INTO THE TERMINAL

STEP 2
MAKE A HALF-TURN WRAP
WITH THE NEEDLE-NOSE PLIERS

STEP 3
MAKE A SECOND HALF-TURN WRAP
WITH THE NEEDLE-NOSE PLIERS

STEP 4
SQUEEZE THE WIRE FLAT
AGAINST THE TERMINAL

Wire-to-Terminal Connections

Make sure the terminal is clean before connecting wires to it.

A terminal that has never been used will be ready for the splicing operation. But if the terminal is not shiny and clean, you should clean it with sandpaper or a sharp scraping tool. Do not use steel wool.

If the terminal has been used before and is clogged or covered with a thick layer of solder, heat the terminal with your soldering iron and wipe away the molten solder with a small rag or wire brush. Take care to prevent bits of molten solder from splattering inside the equipment and on yourself.

Follow the steps shown in *Figure 7-4.* If you are connecting two or more wires to the same terminal, treat each one separately if possible. After making the connections, tug at them to make sure they are secure, and trim away any excess length.

Leave a bit more than 1/16 inch of bare wire between the insulation and the connection.

Now that you have had a chance to see how you are supposed to connect a wire to a terminal, you are in a better position to understand how much insulation you are to strip away. When you have completed the mechanical connections, the insulation on each piece of wire should be a bit more than 1/16 inch away from the connection. Leaving too much bare wire risks short circuits; leaving too little risks melting some of the insulation into the solder joint. It is better to strip away too much insulation than too little. You can aways trim away the excess length after you have made the mechanical connections.

Printed-Circuit Connection

You have just learned how to make the mechanical connections between wires and a terminal. Another kind of connection frequently used in modern electronics is a printed-circuit connection. Printed-circuit boards provide a convenient means for mounting small electronic components. Generally speaking, a printed-circuit board replaces wires with narrow strips of copper foil (conductor) on a board made of an insulating material such as fiberboard or fiberglass.

Figure 7-5 illustrates the procedure for securing components to a printed-circuit board. The component is mounted on the side of the board opposite the side having the copper foil. The leads on the component are inserted through the holes provided for that purpose; the component is pulled firmly against the board; and the leads are bent along the foil pattern on the copper side. Diagonal pliers are used to clip off the excess wire. The finished connection should hold the component securely in place.

**Figure 7-5.
Connecting
Components to a
Circuit Board**

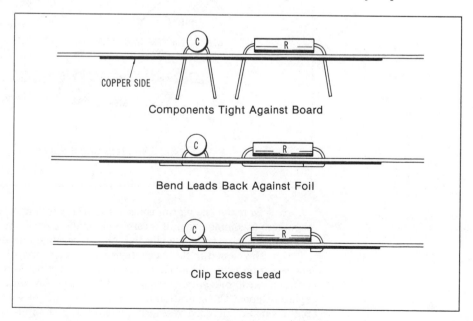

COPPER SIDE

Components Tight Against Board

Bend Leads Back Against Foil

Clip Excess Lead

Wire-to-Wire Connections

A third kind of connection is a wire-to-wire splice. Although splices are necessary for home wiring, they should be avoided wherever possible in electronics work. The two kinds of splices shown in *Figure 7-6*, the pigtail and Western Union, should be used only for temporary connections or under emergency conditions. The high-performance characteristics of modern computer and communications equipment preclude the use of even the best splices on a permanent basis.

**Figure 7-6.
Common Wire Splices**

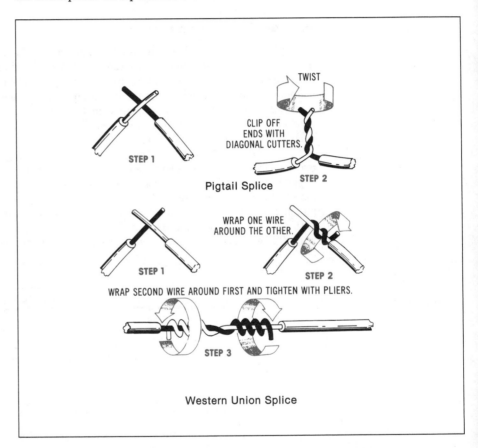

Pigtail Splice

Western Union Splice

To make the pigtail splice, cross the wires and twist both of them together. A common sign of a carelessly made pigtail splice is one wire remaining fairly straight while the other is wrapped around it.

Developed in the early days of the telegraph, the Western Union splice can be neat and mechanically secure. Starting from the crossed-wire position, wrap one wire neatly around the other (as with a poor pigtail splice). Repeat the operation for the other section of wire. (Two poor pigtails make one good Western Union.) Wrap the finished splice with electrical tape.

THE PROPER SOLDERING TECHNIQUE

Soldering sounds like a simple procedure. And it is. But soldering is one of those procedures that has a hundred ways for doing it wrong for every one that is right. We have already described several important points: you must use a 60/40 rosin-core solder, the conductors to be included in the solder joint must be clean and clear of any insulation, and the mechanical connection must be secure.

Preparing the Soldering Iron

Assuming that you've cleaned the points of connection and made a firm mechanical bond, it is time to pick up the soldering iron; however, there are some things you must do before touching the iron to the connection.

First, make sure that the iron is heated to its full temperature; using a lukewarm soldering iron guarantees a poor solder connection. The best way to check the temperature of the iron is to touch your solder to it. If the solder melts against the tip within a few seconds, the temperature and type of solder are correct.

Second, make sure that the soldering-iron tip is properly prepared for the operation. Modern soldering-iron tips—those intended for electronics work—are plated with a material that ensures even distribution of heat and chemically reacts with the solder to improve the electrical qualities of the joint. You know that a tip is properly prepared when it is clean and free of jagged edges and corners.

Remove the oxidation and excess solder from a hot soldering-iron tip by wiping it against a clean, damp cloth or sponge. If solder forms into beads on a hot soldering iron tip, and careful cleaning fails to clear up the problem, replace the tip with a new one.

Oxidation is bound to gather on a hot soldering iron that is waiting for use. The best way to clean away the oxidation and any excess solder is to wipe the tip with a damp cloth or sponge. When you have done that, the tip should remain shiny for several minutes at a time.

When you test the temperature by touching some solder to the tip, the solder should spread quickly and evenly over the surface of the tip. This is called "tinning" the tip. The tip is not properly prepared if the solder forms a ball and rolls off. If that happens, try swirling the hot tip around in a bit of paste rosin for a few seconds, then wipe it clean with a damp cloth or sponge. Try the solder test again. If the solder still beads up on the tip, the plating material is exhausted or you are using a low-quality tip. In any case, it is time to replace the tip with a good one.

Do not scrape modern soldering-iron tips with a file nor scrub them with abrasive materials.

Never scrape modern soldering-iron tips with a file nor scrub them with abrasive materials. Such procedures are more appropriate for heavy-duty soldering irons used in other industries.

Soldering Terminals and Splices

A good solder joint has just enough solder to form a firm and reliable electrical bond between all the parts involved in the connection. The fact that the solder enhances the mechanical bond is secondary. A good solder connection should have a smooth, semi-shiny appearance; the solder should be blended around the shapes of the wire and terminal and should fill in any gaps. Any other appearance is a sign of a poor solder connection.

When you are ready to solder a connection, first touch the soldering iron to the connection. After a few seconds—when you think the connection is hot enough to melt the solder—touch the solder to the connection as shown in *Figure 7-7*. Within a couple of seconds, the solder should melt. And as you gradually feed more solder into the joint, it blends all the way around the connection. Remove the length of solder but leave the tip of the iron applied to the joint for a few more seconds. This provides the heat that is necessary for boiling away the excess rosin.

**Figure 7-7.
Soldering to a Terminal**

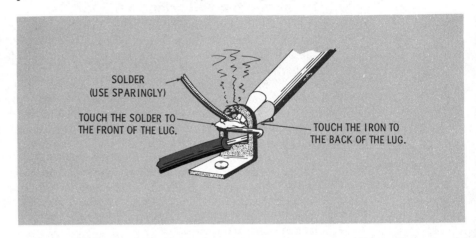

The primary purpose of the soldering iron is to heat the connection to be soldered. The iron is not supposed to melt the solder directly. Once you have properly heated the connection, the solder should melt easily on it, spreading evenly and quickly over all parts of the connection. That is the only way to get a proper solder joint.

Use the soldering iron to heat the connection; use the heated connection to melt the solder. Do not melt the solder directly with the tip of the iron.

It is natural for beginners to try dribbling or painting hot solder onto a connection. Hot solder striking a cold connection causes something known as a "cold-solder joint." This kind of joint has a dull appearance and a rough, uneven texture.

If you are using the soldering iron and solder properly, there should be little doubt about how much solder to use. You should use just enough to cover all parts of the joint and fill in the gaps between the parts.

Do not allow any part of the connection to move during the soldering procedure and while the solder is cooling.

It is important that none of the parts of the connection move while you are doing the soldering and while the solder is cooling. This is one reason why you should make a firm mechanical connection before starting the soldering operation.

If you are using an iron that has a wattage rating appropriate for the kind of work you are doing, you can use a bit of commonsense care to avoid overheating delicate components in the circuit. Heat travels rapidly along wires; so if your iron is too hot for the job at hand, you run the risk of melting away insulation or destroying small parts. To prevent this, use the needle-nose pliers as a heat sink as shown in *Figure 7-8*. The idea is to use the steel in the pliers to carry away heat that would otherwise travel away from the connection to other components.

**Figure 7-8.
Using a Heat Sink**

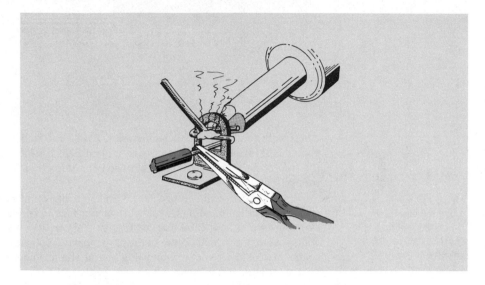

The figure shows that you have to manipulate three items at the same time: soldering iron, solder, and the heat-sink pliers. Unless you happen to have three hands or can hold the solder in your teeth, you will find it convenient to clamp the pliers in place by snapping a rubber band around the handles.

Soldering Printed-Circuit Boards

Areas where the leads of mounted components come through the copper side of a printed circuit board are called "mounting pads." Assuming that you have mounted the component as described earlier in this chapter, all you have to do is flow hot solder over the wire and its pad.

Use the soldering iron to heat the mounting pad and wire; use the heated pad and wire to melt the solder.

Hold a hot, tinned soldering iron tip firmly against the copper pad and wire. Wait a few seconds, then apply the solder to the opposite side of the pad and wire. The heat from the wire and pad should be sufficient to melt the solder and cause it to flow evenly over the entire connection area. See the example in *Figure 7-9*.

Applying too much heat to a mounting pad causes the foil to pull loose from the board.

The worst thing that can go wrong when soldering components to a printed-circuit board is the overheating of the copper foil. Overheating the foil loosens it from the board and destroys the effectiveness of the connection. The entire soldering operation should take no longer than about four seconds. If you find it is taking a significantly longer time, you are running the risk of overheating the foil.

There are a couple of ways to speed up your circuit-board soldering technique. One is to make sure the tip of the soldering iron is operating at full heat and properly prepared (smooth, clean, and tinned).

Figure 7-9.
Solder on a Printed-
Circuit Mounting Pad

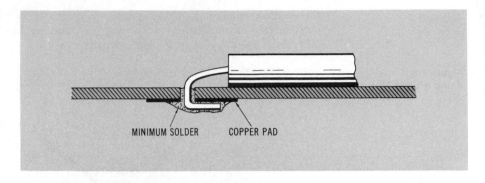

MINIMUM SOLDER   COPPER PAD

Another is to use the smallest-diameter, rosin-core solder you can find. And you should make sure that you have a good heat contact between the tip of the iron and the connection and between the unmelted solder and the connection.

Another common problem with soldering on printed-circuit boards is the application of so much solder that you form a conductive bridge between two different foil pads or tracks. This creates a short circuit that will prevent the circuit from operating properly. You can usually see a solder bridge and remove it by using one of the unsoldering techniques described later in this chapter.

Using too much solder on a printed-circuit connection risks building a short-circuit bridge between foil conductors.

Tinning Stranded Wire

Most of the wire referred to in this book is solid wire. Solid wire has a single conductor covered with vinyl insulation. The only problem with solid wire is that it is not very flexible when compared with "stranded wire." Stranded wire is technically a single conductor, but it is made up of a number of strands of finer wire. Stranded wire is most often used in bundles that run between parts of electronic equipment. These bundles can be bent and repositioned more easily than bundles of solid-conductor wire.

Soldering stranded wire poses a special problem, however. Unless you prepare stranded wire properly, you will find that the strands wander all over the place when you try to connect them to a terminal or fasten them flush against a mounting pad on a printed-circuit board.

Preparing stranded wire for soldering is called tinning, a term with which you are already familiar. You begin by stripping the insulation as shown in *Figure 7-10*. Then twist the strands tightly together. Finally, touch a hot soldering iron on one side of the strands; when they are hot, touch unmelted solder to the other side. If you are doing the job properly, the solder will suddenly flow around and between all the strands, holding them together in a single unit.

After the tinned strands have cooled, shape the wire to fit the terminal or feed it through the printed-circuit board. You will know you have used too much solder if you have difficulty shaping the strands for fitting them through the hole. Remedy that situation by reheating the strands and wiping away the excess solder.

Figure 7-10.
Tinning Stranded Wire

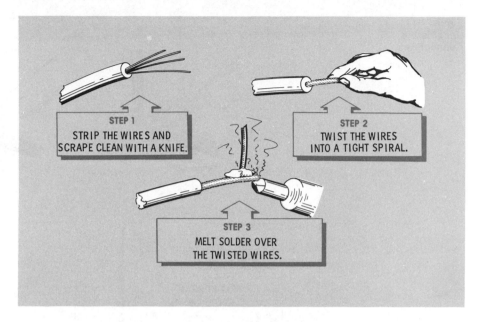

STEP 1
STRIP THE WIRES AND
SCRAPE CLEAN WITH A KNIFE.

STEP 2
TWIST THE WIRES
INTO A TIGHT SPIRAL.

STEP 3
MELT SOLDER OVER
THE TWISTED WIRES.

Summary of Soldering Rules and Precautions

Soldering is a skill, but it is one that can be mastered with practice. Just bear in mind the following rules and precautions:

1. Use only 60/40 rosin-core solder.
2. Make good mechanical connections before soldering.
3. Never use an iron that isn't fully heated.
4. Never use a dirty or worn-out soldering-iron tip.
5. Use the soldering iron to heat the connection; use the heated connection to melt the solder.
6. Use just enough solder to cover all parts of the connection and fill any gaps.
7. Continue applying heat until all rosin is burned away.

UNSOLDERING A CONNECTION

It is often necessary to disconnect a wire or component from a soldered connection. The general idea is to heat the solder connection until the solder melts and, while the solder is still hot, to remove the wire or component from the connection. *Figure 7-11* illustrates the unsoldering procedure for wires and terminals.

Printed-Circuit Board

Solder should be removed from a printed-circuit connection before attempting to pull away the wire or component.

Unsoldering components from a printed-circuit board uses the same general procedure but often requires more care. Too much heat or too much force applied to the component being removed can loosen or tear the copper foil. The recommended procedure begins by removing the solder from the connection before attempting to remove the component.

**Figure 7-11.
Unsoldering
Connections**

STEP 1

HEAT JOINT UNTIL
SOLDER MELTS.

STEP 2

UNWIND WIRE.

STEP 3

PULL WIRE
OUT OF TERMINAL.

There are several ways to go about removing the solder from a printed-circuit connection. One way is to heat the connection to melt the solder, then to carefully wipe away the molten solder with a steel-bristle brush (not steel wool).

Another method is to turn the board upside down to let the molten solder run down onto the hot tip of the soldering iron. The solder you are removing is thus confined to the tip of the soldering iron; and you can remove it by wiping the iron with a damp cloth or sponge.

A more suitable procedure for removing solder from a printed-circuit connection takes advantage of the fact that braided wire can "wick" molten solder away from the connection. A ribbon-like wicking material made of fine, braided wire is commercially available for the purpose. To work properly, the wick must be as hot as the molten solder. Apply the tip of this wicking ribbon to the connection to be unsoldered, then press the tip of a hot soldering iron onto the wicking material. The iron thus heats the wick, and the wick heats the solder connection. Within a few seconds, you will see the solder melt away from the connection and run up into the wicking material. Remove the iron and wicking material to expose the unsoldered connection. Use wire cutters to trim away the solder-saturated tip of the wicking ribbon before using it again.

The fastest, simplest, and most effective desoldering method is the use of a vacuum system that sucks the solder away from the terminal and completely off the printed-circuit board. As illustrated in *Figure 7-12*, you depress a bulb while heating the connection. When the solder melts, you release the bulb to pull the solder away from the connection.

**Figure 7-12.
Simple Vacuum System
for Unsoldering**

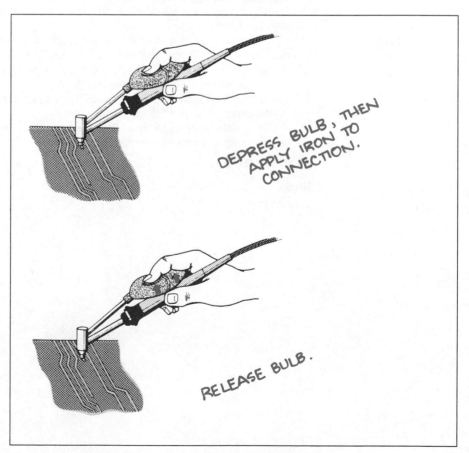

WHAT HAVE WE LEARNED?

1. Solder on an electrical connection prevents oxidation of the metals and improves the mechanical strength of the connection.
2. Use only 60/40 solder for electronics applications. This is a mixture of 60% lead and 40% tin.
3. Rosin prevents oxidation during the soldering process and removes small amounts of foreign materials from the solder connection.
4. Never use acid-core solder for electrical applications.
5. Soldering irons rated between 10 and 75 W are best for most electronics applications.
6. The mechanical connection, not the solder, is supposed to hold the wires and terminals together.
7. Never use steel wool to clean electrical connections.
8. Leave a bit more than $\frac{1}{16}$ inch of bare wire between the insulation and the connection.
9. Remove the oxidation and excess solder from a hot soldering-iron tip by wiping it against a clean, damp cloth or sponge.
10. Do not scrape modern soldering-iron tips with a file nor scrub them with abrasive materials.
11. Use the soldering iron to heat the connection; use the heated connection to melt the solder. Do not melt the solder directly with the tip of the iron.
12. Do not allow any part of the connection to move during the soldering procedure and while the solder is cooling.
13. Using too much solder on a printed-circuit connection risks building a short-circuit bridge between foil conductors.
14. Solder should be removed from a printed-circuit connection before attempting to pull away the wire or component.

KEY WORDS

Rosin
Solder

Quiz For Chapter 7

1. The main purpose of solder in electronics work is to:
   a. make a good mechanical connection.
   b. make and maintain good electrical connections.
   c. hold all the parts together.

2. The principal components of solder are:
   a. lead only.
   b. bismuth and lead.
   c. lead and rosin.
   d. lead and tin.

3. Which one of the following kinds of solder is best for electronics work?
   a. 60/40 rosin-core solder.
   b. 60/40 acid-core solder.
   c. 50/50 rosin-core solder.
   d. 40/60 solid-core solder.

4. Which one of the following power ratings for soldering irons is best for most kinds of electronics work?
   a. 5 W.
   b. 50 W.
   c. 500 W.
   d. 5000 W.

5. What is the proper procedure for tinning the tip of a soldering iron?
   a. File or scrape away foreign material while the tip of the iron is cold. Heat the iron and melt some solder on it.
   b. Melt some solder on the tip of the iron, then hold the hot tip in cool, running water.
   c. Wipe the hot tip clean with a damp cloth or sponge, then melt a bit of solder on the tip and wipe away the excess.

   d. Melt some fresh, acid-core solder on the hot tip of the iron, then wipe away the excess solder on your pants or shirt sleeve.

6. Which one of the following is the best soldering technique?
   a. Use the iron to melt fresh solder onto the connection. When the solder cools, heat the connection with the iron to melt the solder again.
   b. Use the iron to heat the connection; use the hot connection to melt the solder.
   c. Use the iron to melt the solder; let the hot solder heat the connection.

7. Gently wiggling the component while its solder connection is cooling:
   a. ensures a good electrical and mechanical connection.
   b. works any foreign materials away from the connection.
   c. helps the solder cool faster.
   d. guarantees a poor solder connection.

8. Which one of the following suggestions is an acceptable way to remove solder from a connection you are unsoldering?
   a. Melt the solder with a hot soldering iron, and use a simple vacuum tool to suck up the molten solder.
   b. Melt the solder with a hot soldering iron, and use a simple vacuum tool to blow the molten solder away from the connection.
   c. Hold a hot soldering iron on the connection until the solder burns away.

Understanding Inductors and Transformers

ABOUT THIS CHAPTER

This chapter describes two kinds of wirewound devices: inductors and transformers. You will learn how these devices operate and how they are used for controlling ac electrical power. When you complete your study of this chapter, you should be able to identify various kinds of inductors and transformers, and you should have a good understanding of how they are rated.

COMMON CHARACTERISTICS OF INDUCTORS AND TRANSFORMERS

Inductors and transformers work according to the same basic principle of electromagnetism. You learned in Chapter 1 that current flowing through a wire creates a magnetic field. This is one of the principles that inductors and transformers have in common. They also share a principle that says it is possible for a moving magnetic field to generate, or induce, a voltage in a conductor that is immersed in that field.

Generating a Magnetic Field with a Current

The strength of a magnetic field produced by current flowing through a coil of wire depends partly on the amount of current flowing through the coil and the number of turns of wire in the coil.

Inductors and transformers both include a coil of wire that carries a current that creates a magentic field. The strength of a current-generated magnetic field depends on a number of factors. The strength of the magnetic field depends on the amount of current flowing through the coil of wire. As shown in *Figure 8-1*, the larger the amount of current, the stronger its magnetic field.

The strength of a current-generated magnetic field also depends on the number of turns of wire, the diameter and length of the coil, and the kind of material the wire is wrapped around. These factors, however, are fixed at the time the inductor or transformer is constructed; so the only control you have over the strength of a current-generated magnetic field is to control the amount of current applied to the coil.

**Figure 8-1.
Current Flow and
Magnetic-Field Strength**

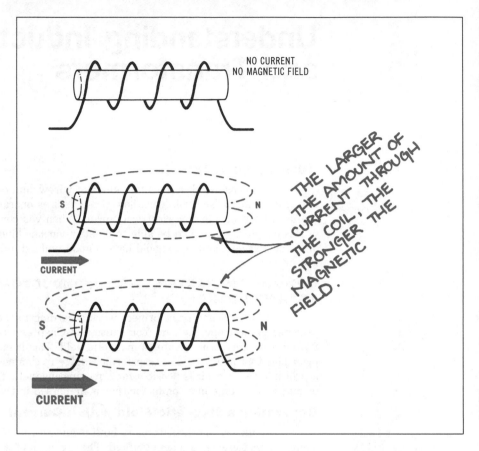

NO CURRENT
NO MAGNETIC FIELD

S   N

CURRENT

S   N

CURRENT

THE LARGER THE AMOUNT OF CURRENT THROUGH THE COIL, THE STRONGER THE MAGNETIC FIELD.

Reversing the direction of current flow through a coil of wire reverses the polarity of its magnetic field.

A current-generated magnetic field, like the magnetic field around a bar or horseshoe magnet, has a north and south pole. The north and south poles for a bar or horseshoe magnet never change. As shown in *Figure 8-2*, however, you can reverse the magnetic polarity of a current-generated magnetic field by reversing the direction of current flow through the coil.

Figure 8-2.
Direction of Current
Flow Determines
Magnetic Polarity

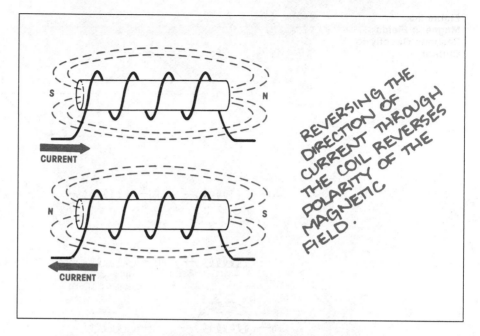

REVERSING THE DIRECTION OF CURRENT THROUGH THE COIL REVERSES POLARITY OF THE MAGNETIC FIELD.

You have just learned that the strength of a magnetic field around a coil of wire is determined by the amount of current flowing through the coil, and the polarity of the magnetic field is determined by the direction of current flow. *Figure 8-3* shows how this magnetic field responds to a typical ac waveform. When the current level is zero, there is no magnetic field. But as the current gradually increases in the positive direction, the strength of the field increases with a particular polarity. Then the current gradually decreases to zero, only to increase to a maximum level in the opposite direction. Notice how the strength and polarity of the magnetic field responds directly to the amount and direction of current flowing through the coil.

**Figure 8-3.
Magnetic Fields
Respond Directly to
Current**

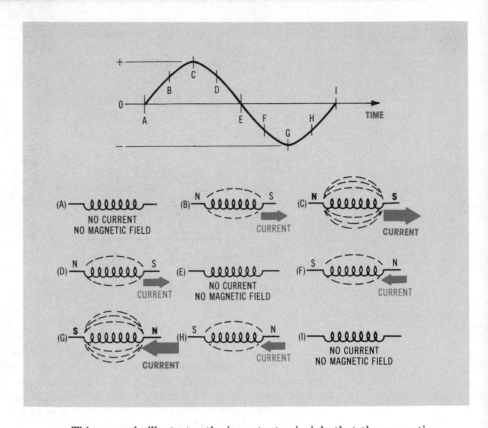

The magnetic field around
a conductor expands and
collapses as current
through the conductor in-
creases and decreases.

This example illustrates the important principle that the magnetic
field surrounding a conductor expands and collapses and changes magnetic
polarity in response to changing amounts and directions of current flowing
through the coil. Inductors and transformers both take advantage of these
principles.

Generating a Voltage with a Magnetic Field

A fluctuating magnetic
field can generate a voltage in a conductor that is
within the field.

Inductors and transformers share one other important principle: a
moving magnetic field induces a voltage in a conductor that is immersed in
that moving field. You know that applying ac current to a coil of wire
produces a fluctuating magnetic field. As shown in *Figure 8-4*, a voltage is
generated in a wire that is in the path of this field.

The amount of induced voltage depends on a number of factors,
but most are of no real concern when you are using ready-made inductors
and transformers. You do have control over two important factors,
however: the strength of the magnetic field and the rate at which it is
moving.

**Figure 8-4.
Inducing Voltage In a
Conductor**

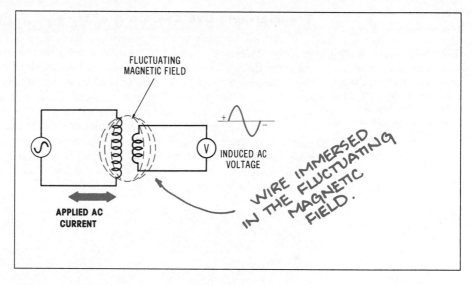

FLUCTUATING
MAGNETIC FIELD

INDUCED AC
VOLTAGE

WIRE IMMERSED IN THE FLUCTUATING MAGNETIC FIELD.

APPLIED AC
CURRENT

The amount of voltage
magnetically induced in a
conductor is proportional
to the strength of the
magnetic field and the
rate at which magnetic
field is changing.

The amount of voltage that is magnetically induced in a conductor
depends partly on the strength of the magnetic field. The stronger the
magnetic field, the greater the amount of voltage induced in the conductor.
The amount of voltage induced in a conductor also depends on how fast the
magnetic field is growing or collapsing. The faster the field is changing, the
larger the amount of voltage induced in the conductor. What happens when
the magnetic field is not changing at all? There can be no voltage induced
in the conductor, no matter how strong that stationary magnetic field
might be.

The polarity of a voltage
that is magnetically in-
duced in a conductor de-
pends on the polarity of
the magnetic field and the
direction in which the
field is changing.

The polarity of the voltage induced in a conductor also depends on
two factors: the polarity of the magnetic field (north or south) and the
direction in which it is changing (expanding or collapsing).

DIFFERENCES BETWEEN INDUCTORS AND TRANSFORMERS

Inductors and transformers both use the two principles just
described: current flowing through a conductor produces a magnetic field
and a moving magnetic field can induce a voltage in a conductor. The two
devices handle the principles in a somewhat different fashion, however.

Transformers Use Two Coils of Wire

Transformers use two
coils, the primary winding
and secondary winding.
Fluctuating current is applied to the primary winding, and the induced
voltage is taken from the
secondary winding.

Transformers use two different coils of wire. A fluctuating current is applied to one of the coils to create a moving magnetic field. This moving magnetic field induces a voltage in the second coil of wire. *Figure 8-5* shows the arrangement of coils for a basic transformer. Current is applied to a coil that is called the "primary winding." The resulting magnetic field induces a voltage in the coil called the "secondary winding."

Transformers are used for transforming current and voltage from one level to another. One level of current and voltage is applied to the primary winding and another is taken from the secondary winding. Later discussions in this chapter describe practical transformer specifications and applications devices in greater detail.

**Figure 8-5.
A Basic Transformer**

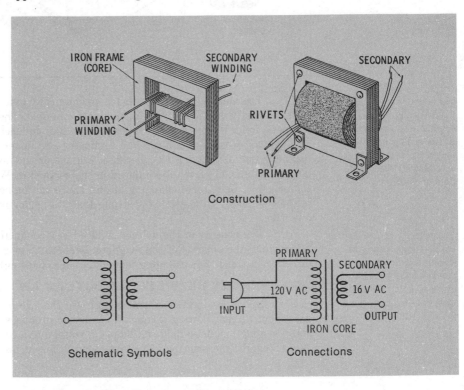

Inductors Use a Single Coil of Wire

An inductor uses a single
coil of wire. A voltage is
magnetically induced in
the same coil that is carrying the magnetizing
current.

Inductors work according to the same general principles as transformers but with one important difference: inductors are made of a single coil of wire. Voltage is magnetically induced in the same coil of wire that carries the magnetizing current. *Figure 8-6* shows the action of a basic inductor. As you increase the amount of current applied to the coil, it creates an expanding magnetic field. This expanding field induces a voltage in the very same coil of wire. As you decrease the amount of current applied to the coil, the magnetic field contracts and induces a voltage of the opposite polarity in its winding.

**Figure 8-6.
Basic Inductor**

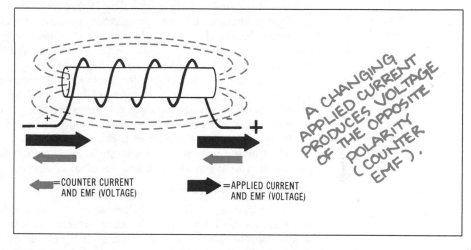

=COUNTER CURRENT
AND EMF (VOLTAGE)

=APPLIED CURRENT
AND EMF (VOLTAGE)

A CHANGING APPLIED CURRENT PRODUCES VOLTAGE OF THE OPPOSITE POLARITY (COUNTER EMF).

Counter EMF

The counter emf induced in the coil of an inductor has a polarity that tends to counter any change in the direction of magnetizing current.

The voltage that is self-induced in the coil of an inductor always has a polarity that tends to oppose any change in the amount of magnetizing current. For this reason, the self-induced voltage is called the "counter emf." ("Emf" stands for electromotive force, another expression for voltage.) While you are increasing the amount of current applied to an inductor, the counter emf has a polarity that tends to counter your increase in current. Likewise, while you are decreasing the amount of current applied to an inductor, the counter emf takes on the polarity that tends to counter the decrease in current flow.

The counter emf in an inductor is generated by a changing current applied to the coil of wire. As long as the applied current is changing, counter emf is present to resist the change. Counter emf is never sufficient to cut off the magnetizing current, but it does react against changes in the amount of current.

Whenever the current applied to an inductor is constantly changing, the inductor constantly opposes the flow of this current. This opposition to current flow is called inductive reactance.

Inductors thus oppose changes in current flow. As long as the current applied to an inductor is changing, the inductor opposes the current. This reaction against a constantly changing current is called "inductive reactance."

Inductors are mainly used for controlling the amount of current flowing in ac circuits. You will learn more about practical inductors and their applications in the next section of this chapter.

INDUCTORS

An inductor offers opposition to the flow of ac current. The amount of opposition depends on the value of the inductance and the rate at which that current is attempting to change.

You have already learned that an inductor is basically a device made up of a single coil of wire. Current applied to this coil generates a magnetic field, and as long as this field is changing, the inductor opposes any further change. In short, an inductor opposes any change in the amount of current flowing through it. The amount of opposition to a change in current flow depends on the value of the inductance and the rate at which the current is attempting to change.

An inductor has little effect on the performance of a dc circuit, no matter how large its inductance value might be. The reason is that a dc circuit normally shows a change in current when the power is first applied, and then again when power is removed. Inductors in such instances do their work only while the current is changing.

Current is changing constantly in an ac circuit. So inductors have a great deal of influence on the operation of an ac circuit.

Values of Inductance

The basic unit of inductance is the henry.

Figure 8-7 shows that inductors are available in a variety of forms and sizes. Even so, they are all basically nothing more than coils of wire wrapped around some sort of core material. The basic unit of inductance is the "henry" (abbreviated H). Some of the largest inductors, those using heavy iron core materials, have inductance ratings on the order of 1 to 10 H.

**Figure 8-7.
Inductors**

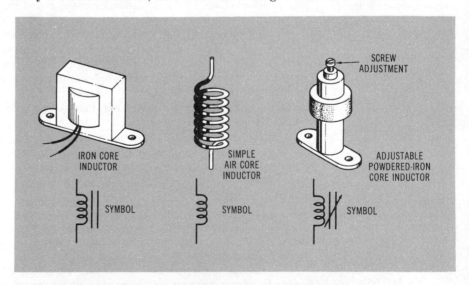

Smaller inductors have a diameter about the same as that of a ballpoint pen and have soft, powdered-iron cores. These inductors have values on the order of a few hundred millihenrys (one millihenry is one-thousandth of a henry). Inductors constructed in this fashion are sometimes adjustable. The soft, powdered-iron cores can be screwed in and out of the coil. The greater the amount of core material within the coil of wire, the greater the value of inductance.

The smallest-valued inductors are simply several turns of wire. The core material is nothing more than the air that surrounds the windings. These inductors have values on the order of several hundred microhenrys (one microhenry is equal to one-millionth of a henry).

Inductive Reactance

The amount of inductive reactance in an ac circuit depends on the value of the inductance and the frequency of the applied current.

Inductors used in ac circuits offer a constant opposition to current flow called inductive reactance. The amount of inductive reactance depends partly on the value of the inductance: the larger the value of the inductance, the larger the amount of opposition to ac current. The amount of inductive reactance also depends on the frequency of the current—the number of times the current switches polarity each second. The higher the applied frequency, the greater the amount of inductive reactance.

Current Rating for Large Inductors

You know that inductors are rated according to their inductance values. The larger versions are also rated according to the maximum amount of current that can safely pass through their coil of wire. A typical inductor of this type might be rated as 4 H, 10 A. This means it has a value of 4 H and that you can use it in a circuit that passes no more than 10 A of current. Exceeding the current rating risks overheating the device and burning open the wiring.

Resistance and Impedance for Small Inductors

Small inductors have a large number of turns of very fine wire. This wire has a resistance that you can measure with an ohmmeter. The resistance of the inductor does not change with the applied frequency.

The combination of inductive reactance and wire resistance in an inductor is called impedance. Impedance is measured in ohms.

Smaller inductors thus offer two different kinds of opposition to current flow through them—inductive reactance and resistance. This combination of oppositions to current flow is called "impedance." Inductors intended for high-frequency applications are often specified according to the amount of impedance they offer at a certain frequency, rather than according to their value of inductance.

TRANSFORMERS

You have already learned that a transformer uses two separate windings. Ac power is applied to the primary winding. The current flowing in this winding generates a magnetic field that expands and collapses in response to the current flowing through it. The changing magnetic field induces a voltage in the secondary winding.

Transformers convert ac voltages and currents from one level to another.

A "transformer" is an electrical device that converts ac voltages and currents from one value to another. Transformers are available in a number of varieties and sizes. Door bell transformers, for example, reduce the 120 V ac that is available in your home to safer levels of 6 or 12 V ac. A different kind of transformer in your television set steps up 40 V ac to 20,000 V or more for operating the picture tube.

Figure 8-8 shows how transformers are used in the plans for delivering electric power to your home. Electric power generators operated by your local power utility company produce voltages on the order of 4000 V ac. Transformers step up that voltage to about 400,000 V for transmission across long distances. When the power lines reach your city or town, a power substation uses transformers to step down the voltage to about 440 V ac for distribution to the neighborhoods. The familiar transformer located on a utility pole near your home steps down the voltage to 240 and 120 V ac for direct use in your home.

**Figure 8-8.
Transformers in a
Power-Distribution
System**

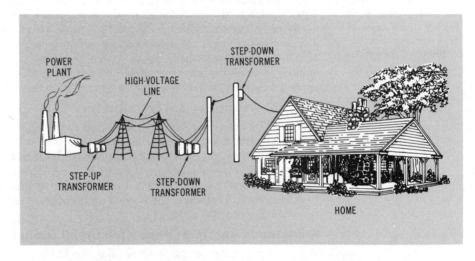

Transformers cannot change the level of dc voltages and currents.

A step-up transformer increases an ac voltage level. A step-down transformer decreases the applied ac voltage level.

Transformers step up and step down ac voltages. They also step up and step down the amounts of available ac current. Transformers cannot step up or step down dc currents and voltages.

Figure 8-9 is a schematic diagram of a door bell transformer circuit. You can see that a plug is used to connect the primary winding of the transformer to 120 V ac. The secondary winding then provides 12 V ac for use in lower-voltage circuits. This is an example of a step-down transformer. A "step-down transformer" is one that reduces the applied ac voltage level. A "step-up transformer" increases the ac voltage level.

CHARACTERISTICS OF TRANSFORMERS

Now that you understand the fundamental principles of transformer action, you are ready to learn more about practical transformers and the way they are rated.

A Basic Transformer Circuit

Figure 8-10 is a schematic diagram for a basic transformer circuit. This circuit demonstrates the principles and characteristics of power transformers. If you decide to build this circuit, you can purchase an inexpensive power transformer called a filament transformer from most electronics supply stores. As shown in the figure, most filament transformers step down 120 V ac to 12.6 V ac.

Figure 8-9.
Step-Down Transformer

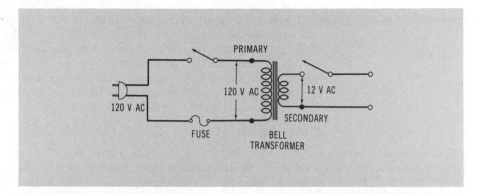

Figure 8-10.
Power-Transformer
Circuit

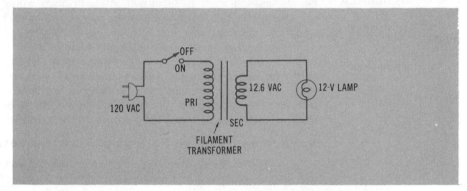

When you insert the plug into an ordinary outlet, you apply 120 V ac to the primary winding of the transformer. As a result, you should immediately find about 12 V ac at the terminals of the secondary winding.

As long as there is no electrical device connected to the secondary winding, little current flows through the primary winding and none through the secondary winding. Connect the terminals of the secondary winding to a 12-V lamp, however, and current will flow in both the primary and secondary windings.

Voltage Ratio

Because a step-up transformer increases ac voltage levels, and a step-down transformer decreases ac voltage levels, you can expect the secondary voltage for a step-up transformer to be greater than the voltage applied to its primary winding. By the same token, you can expect the secondary voltage for a step-down transformer to be less than the voltage at the primary winding. The filament transformer in *Figure 8-10* is a step-down transformer because it steps down the 120 V ac applied to its primary winding to 12 V ac.

The voltage ratio of a transformer is the ratio of its primary voltage to its secondary voltage.

One of the most important ratings for a transformer is its "voltage ratio." The voltage ratio for a transformer is the ratio of its primary voltage to its secondary voltage. Suppose that you are working with a transformer that has a voltage ratio of 10 to 1 (often written as 10:1 or

10/1). This means every 10 V applied to the primary winding will produce just 1 volt at the secondary winding. Apply 50 V ac to the primary winding, for instance, and you will find 5 V at the secondary winding.

What is the voltage ratio for a transformer that produces 12 V at its secondary when 120 V is applied to its primary winding? The ratio of primary voltage to secondary voltage is 120:12, or 10:1. Most filament transformers have a 10:1 step-down voltage ratio.

Turns Ratio

The turns ratio of a transformer is the ratio of the number of turns of wire in its primary winding to the number of turns of wire in its secondary winding. The turns ratio of a transformer is identical to its voltage ratio.

The "turns ratio" of a transformer is the ratio of the number of turns of wire in the primary winding to the number of turns of wire in its secondary winding. A transformer that has a 6:1 turns ratio, for instance, has six times as many turns of wire in the primary winding as in the secondary winding. If this transformer happens to have 600 turns of wire in its primary, it must have just 100 in its secondary.

The turns ratio of a transformer is directly related to its voltage ratio. In fact, the two ratios are exactly the same. So if you are working with a transformer that has a 10:1 voltage ratio, you can bet that there are ten times as many windings in the primary as in the secondary. This transformer might have a thousand turns of wire in its primary. If that is the case, there has to be a hundred turns in the secondary winding.

Voltage and Current Ratings of Transformers

Power transformers are rated according to their voltage ratios and the amount of power (the product of voltage times current) that their windings can handle. Power transformers do run hotter than most other kinds of electronic components. An iron-core transformer that is operating just within its power rating might feel very hot to the touch. But you know you have exceeded the power rating if you can sense the odor of hot wiring or, worse, see some smoke rising from it.

A typical filament transformer might be specified as having a 120-V primary and a 12.6-V secondary. It might also show a maximum power rating of 12 W. The current rating is equal to the power rating divided by the voltage specification for the primary winding. So a transformer that has a power rating of 12 W and a primary voltage specification of 120 V can handle up to 12/120, or 0.1 A of current. Some manufacturers save you the trouble of doing the arithmetic and directly specify the maximum current rating of the windings.

AUDIO TRANSFORMERS

Audio transformers can be used for changing voltage and current levels, but their primary purpose is to match the impedance of one part of a circuit with the impedance of a different part of a circuit.

Most loudspeakers have an impedance of 8 Ω or, in some instance, 16 Ω. Electronic amplifiers—electronic circuits that boost and control the power level of the sound—have much higher output impedances. For the sake of discussion, let's say a typical audio amplifier has an output impedance of 4000 Ω.

If you were to connect a circuit having an output impedance of 4000 Ω with a loudspeaker having an impedance of 8 Ω, the mismatch would be so bad that you would most likely hear no sound at all. In fact, you might burn out the output section of the amplifier.

Fortunately, the proper audio transformers are usually built into the amplifier circuit. The matching is already done for you; you simply make sure you connect the proper speaker—8 Ω or 16 Ω—to the amplifier. The audio transformer has a primary winding with an impedance of 4000 Ω and a secondary of 8 or 16 Ω.

Although audio transformers are usually built into the amplifier unit, you still have an opportunity to mess things up. Suppose you want to connect a second 8-Ω loudspeaker to the same terminals reserved for one. That changes the total loudspeaker impedance to 4 Ω. The audio transformer has an 8-Ω secondary, so you are creating an impedance mismatch that is bound to degrade the quality of the sound and total power output level.

RADIO-FREQUENCY TRANSFORMERS

Like audio transformers, radio-frequency transformers are often used to match the impedance of two processing devices. If you have a standard television receiver connected to a cable television service, you can probably locate a radio-frequency transformer.

The point where you normally connect an antenna to a television set has an impedance of 300 Ω. The cable that feeds programming to your television, however, has an impedance of 75 Ω. It is thus important that you insert a small radio-frequency transformer from the cable connection to the vhf terminals on the television set.

This transformer should be wound in such a way that the primary winding has an impedance of 75 Ω (to the cable) and a secondary winding of 300 Ω (to the vhf terminals on the receiver). *Figure 8-11* shows a typical matching transformer for cable television.

**Figure 8-11.
Cable-TV Matching
Transformer**

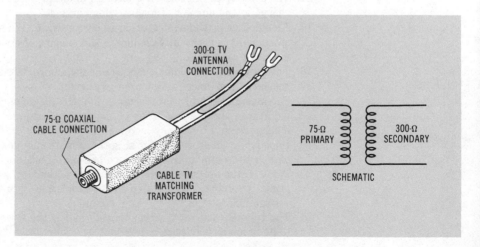

WHAT HAVE WE LEARNED?

1. Current flowing through a wire creates a magnetic field.
2. The larger the amount of current flowing through a coil of wire, the stronger its magnetic field.
3. As in common bar and horseshoe magnets, the magnetic field around a current-carrying coil of wire has a north and south pole.
4. You can change the magnetic polarity of a current-generated magnetic field by reversing the direction of current through the coil.
5. Magnetic fields expand as current increases; the fields collapse as current decreases.
6. A moving magnetic field can induce a voltage in a conductor that is immersed in that field.
7. The amount of voltage magnetically induced in a conductor depends partly on the strength of the magnetic field. The stronger the field, the larger the amount of induced voltage.
8. The amount of voltage magnetically induced in a conductor depends partly on the rate at which the field is changing. The greater the rate of change, the larger the amount of induced voltage.
9. No voltage is induced in a conductor when there is no motion between the conductor and the magnetic field, no matter how strong the magnetic field might be.
10. The polarity of the voltage magnetically induced in a conductor depends on the polarity of the magnetic field and the direction in which the magnetic field is changing.
11. A basic transformer uses two different windings. A fluctuating current is applied to the primary winding, and the induced voltage is taken from the secondary winding.
12. A basic inductor uses a single coil of wire. A fluctuating current applied to the inductor creates a magnetic field that induces a voltage in the same coil.
13. The voltage induced in an inductor is called a counter emf.
14. The polarity of the counter emf in an inductor is such that it opposes any change in the amount of magnetizing current.
15. An inductor provides opposition to current flow. When the current applied to an inductor is constantly changing, the amount of opposition is called inductive reactance.
16. The basic unit of inductance is the henry, abbreviated H.
17. Inductive reactance is measured in ohms.
18. The amount of inductive reactance in a circuit depends on the value of inductance and the frequency of the applied current. Increasing both the inductance and frequency increases the amount of inductive reactance.
19. Impedance, measured in ohms, is the combined opposition to current flow offered by the inductive reactance and wire resistance of an inductor.
20. When a step-up transformer is used, the secondary voltage is larger than the primary voltage. When a step-down transformer is used, the secondary voltage is less than the primary voltage.
21. The voltage ratio for a power transformer is the ratio of its primary voltage to its secondary voltage.

22. The turns ratio for a power transformer is the ratio of the number of turns in its primary winding to the number in its secondary winding.
23. The voltage and turns ratios have the same value for a power transformer.

KEY WORDS

Counter emf
Impedance
Inductive reactance
Inductor
Primary winding
Secondary winding

Step-down transformer
Step-up transformer
Transformer
Turns ratio
Voltage ratio

Quiz for Chapter 8

1. Which one of the following statements is correct?
   a. Increasing the amount of current flowing through a coil of wire increases the strength of its magnetic field.
   b. Changing the amount of current flowing through a coil of wire changes the polarity of its magnetic field.
   c. The strength of a current-generated magnetic field depends partly on the direction of current flow.

2. The polarity of the magnetic field around a current-carrying conductor depends on the:
   a. amount of current flowing through the conductor.
   b. direction of the current flowing through the conductor.
   c. resistance of the conductor.
   d. frequency of the current applied to the conductor.

3. Which one of the following factors is partly responsible for determining the amount of voltage that is magnetically induced in a conductor?
   a. Rate of change of the magnetic field.
   b. Polarity of the magnetic field.
   c. Direction of change of the magnetic field.

4. Which one of the following factors is partly responsible for determining the polarity of voltage that is magnetically induced in a conductor?
   a. Strength of the magnetic field.
   b. Rate of change of the magnetic field.
   c. Direction of change of the magnetic field.

5. The counter emf in an inductor:
   a. always opposes any change in the amount of magnetizing current.
   b. creates the magnetizing current.
   c. always supports any change in the amount of magnetizing current.
   e. eliminates the magnetizing current.

6. The opposition to a constantly changing current though an inductor is called:
   a. resistance.
   b. inductance.
   c. resistive reactance.
   d. magnetic reactance.
   e. inductive reactance.

7. Inductance is measured in:
   a. henrys.
   b. ohms.
   c. ampere-turns.
   d. webers.

8. Inductive reactance is measured in:
   a. henrys.
   b. ohms.
   c. ampere-turns.
   d. webers.

9. Impedance is measured in:
   a. henrys.
   b. ohms.
   c. ampere-turns.
   d. webers.

10. Increasing the frequency of an alternating current applied to an inductor:
   a. increases the amount of inductance.
   b. decreases the amount of inductance.
   c. increases the amount of current flowing.
   d. increases the amount of inductive reactance.
   e. decreases the amount of inductive reactance.

11. A certain power transformer has a voltage ratio of 25:1. What is the secondary voltage when 100 V is applied to the primary winding?
   a. 4 V.
   b. 25 V.
   c. 75 V.
   d. 125 V.
   e. 2500 V.

12. What is the turns ratio for a power transformer that has 800 turns of wire in its secondary winding and 600 turns in its primary?
   a. 1:800.
   b. 1:600.
   c. 3:4.
   d. 4:3.
   e. 600:1.
   f. 800:1.

Understanding Capacitors

ABOUT THIS CHAPTER

A capacitor is another very basic but highly useful circuit component. Because they can regulate current, as do resistors and inductors, capacitors are used for this purpose in most circuits. Upon completing your study of this chapter, you will understand what capacitors are, how they are used, and how they are connected into circuits.

CIRCUIT COMPONENTS

Most electronic circuits include combinations of three basic components: resistors, inductors, and capacitors. Each reacts in a different way to ac and dc voltages and currents. You have already learned a great deal about resistors and inductors in previous chapters. A resistor controls electricity by opposing the flow of current. This reaction is called resistance. Resistance has the same effect on the flow of both ac and dc current.

An inductor controls electricity by regulating the flow of ac current. The magnetic field in an inductor cuts its own windings and, thus, develops a voltage that opposes a change in current. This is called inductance. Inductance has a far greater effect in ac circuits than it does in dc circuits.

A capacitor controls electricity by regulating changes in ac voltage levels. It stores and gives up electrical charges that oppose any change in voltage. This is properly called capacitance.

HOW A CAPACITOR WORKS

A capacitor, sometimes called a condenser, is manufactured in a number of different shapes and sizes. *Figure 9-1* illustrates the most common kinds of capacitors. Several of each kind of capacitor are probably in your home. You can find them in radios, television receivers, audio systems, and just about any other kind of electronic equipment you can imagine. Capacitors are also used with some electric motors and in the electrical system of your car.

A capacitor is made up of two conductors separated by a "dielectric." The conductors in a capacitor are usually plates of metal or strips of metal foil. The dielectric is an insulating material. *Figure 9-2* shows a capacitor that fits this basic definition.

A basic capacitor is made up of two conductors separated by an insulating material called a dielectric.

Figure 9-1.
Types of Capacitors

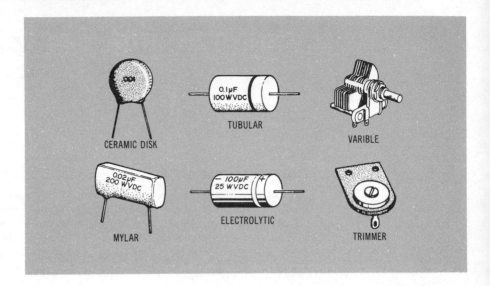

CERAMIC DISK

TUBULAR

VARIBLE

MYLAR

ELECTROLYTIC

TRIMMER

Figure 9-2.
Basic Capacitor
Construction

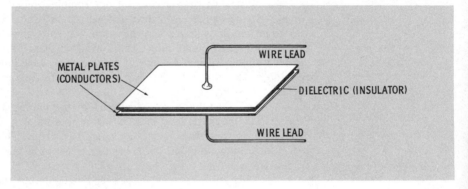

METAL PLATES
(CONDUCTORS)

WIRE LEAD

DIELECTRIC (INSULATOR)

WIRE LEAD

Variations of the basic definition of a capacitor produce the various shapes and types that are available today. For example, the tubular capacitor shown in *Figure 9-3* uses lengths of metal foil as the plates. These foil plates are separated by strips of paper treated to form a good dielectric. Wire leads connected to the exposed ends of the foil provide the electrical connections to the outside. The entire assembly is tightly rolled into a spiral and sealed in a plastic jacket.

The structure of a capacitor follows the fundamental principles of voltage and current as applied to conductors and insulators. Assume, as shown in *Figure 9-4*, that the plates of a capacitor are connected through a switch to a battery. The open switch prevents the battery voltage from being applied to the plates of the capacitor. A voltmeter shows zero volts across the capacitor.

Figure 9-3.
Tubular Capacitor

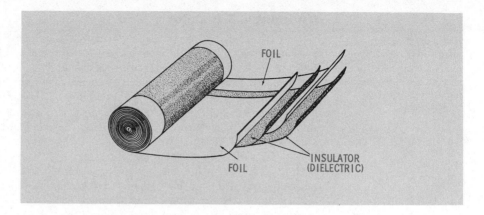

Figure 9-4.
Capacitor with No
Voltage Applied

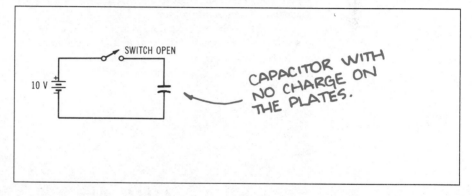

Suppose the switch is now closed. You might think that no current will flow because current cannot flow through a dielectric (insulator). Indeed, current cannot flow through the dielectric; however, some current does flow through the circuit, as shown in *Figure 9-5*.

Figure 9-5.
Charging Current in a
Capacitor Circuit

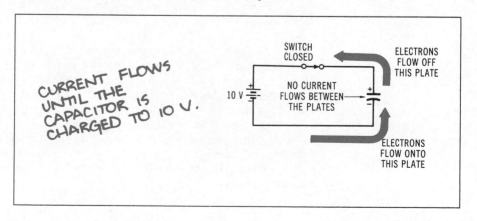

Charging a Capacitor

Electrons flow onto one plate and off the other plate of a capacitor until the voltage between the plates is equal to the source voltage.

Current flows in the capacitor circuit shown in *Figure 9-5* until the voltage across the capacitor is equal to the voltage of the battery. Current flows until the charge on the capacitor is the equivalent of 10 V in this example. All electrons have a negative charge. They are repelled by the negative terminal of the battery and attracted by the positive terminal. The negative terminal of the battery causes electrons to accumulate on the upper plate of the capacitor, and the positive terminal of the battery pulls electrons away from the lower plate.

A deficiency of electrons results in a positive voltage on the bottom plate of the capacitor, and an excess of electrons makes the upper plate negative. Current flows rapidly at first, but more slowly as the voltage across the capacitor builds up to the same voltage as the battery. When the voltage on the capacitor matches the voltage of the battery, current flow stops altogether.

When the current flow stops, a field of force, equal to 10 V in this example, exists between the two plates of the capacitor. This field, called an electrostatic force field, has a direction of force shown in *Figure 9-6*—from negative to positive. The excess electrons are attracted to the positive plate. It is this attraction that develops the force.

**Figure 9-6.
A Charged Capacitor**

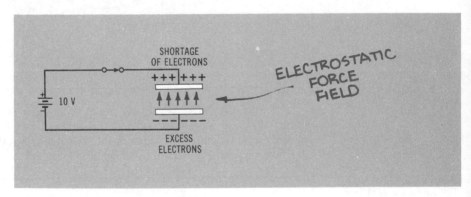

Suppose the switch is now opened. Will the electrostatic force of 10 V disappear? You will find that the voltage across the capacitor is still 10 V, just as it was before you opened the switch. The imbalance of electrons remains because there is no path for current flow that would allow the excess electrons to flow from the negative to the positive plate.

You have just studied an example of the process of charging a capacitor. Charging a capacitor is the process of building up a voltage between the plates—placing an excess of electrons on one plate and creating a deficiency on the other. The capacitor in the preceding example was charged to 10 V.

A charged capacitor can hold its voltage, even after it is disconnected from the circuit.

Some capacitors can be charged to extremely high voltages and will retain their charge for a long time. The high-voltage capacitor in the picture-tube section of your color television set takes on a charge of 20,000 V or more. Equally important is the fact that this capacitor can hold that charge for quite some time after the television is turned off. So be careful when working around high-voltage capacitors, even when you have turned off the equipment. Such capacitors should be discharged before you do any work around them.

Discharge Current

When a capacitor discharges, excess electrons on one plate flow through external connections to the plate that has a shortage of electrons. The discharge is complete when there is no longer any difference in charge between the two plates.

When a capacitor discharges, the excess electrons return to the positive plate. As this happens, the voltage decreases. It decreases rapidly at first, then more slowly as the voltage approaches zero. A capacitor is completely discharged when there is no longer an imbalance between the number of electrons on the two plates.

It is important to realize that current also flows in a capacitor circuit when the capacitor is discharging. As shown in *Figure 9-7*, discharge current flows between the two plates.

Figure 9-7.
Discharging a Capacitor

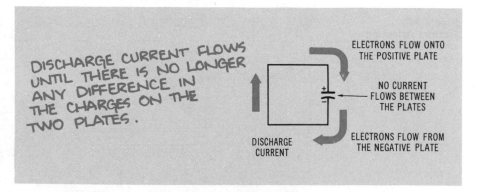

You have a good understanding of the operation of a capacitor circuit when you can see how current flows in a capacitor circuit under two conditions: when the capacitor is charging to a higher voltage level and when it is discharging to a lower voltage level. Current flows in a capacitor circuit only when its voltage is changing.

Current flows in a capacitor circuit only when there is a change in the voltage.

Dc Circuit

In a dc circuit, a capacitor blocks the flow of current.

It is often said that a capacitor blocks dc current. This is quite true if you neglect the fact that the capacitor circuit draws charge current when the dc power is first switched on, and it provides discharge current when the dc power is turned off. Once the dc voltage level is established, current no longer flows through the circuit. In any event, electrons never actually pass through the dielectric material.

Ac Circuit

In an ac circuit, a capacitor allows the voltage to alternately charge and discharge the plates.

In ac circuits, current flows back and forth through a capacitor circuit. Charge current flows through the circuit while the applied voltage is increasing, and discharge current flows through the circuit while the applied voltage is decreasing. But remember that current never flows directly through the capacitor. It flows only through the circuit between the voltage source and the capacitor.

RATING CAPACITORS

Capacitors are rated according to two factors: their capacity for storing electrons and the ability of the dielectric material to withstand the voltage applied to the plates. The capacity of a capacitor is specified in terms of its "capacitance" value. Its ability to withstand the voltage applied to the plates is specified as its voltage rating.

Capacitance Values

A capacitor is rated in terms of its ability to store quantities of electrons. This factor, as illustrated in *Figure 9-8*, is determined by the area of the plates, the distance between the plates, and the type of dielectric material that separates the plates.

**Figure 9-8.
Factors Affecting
Capacitance**

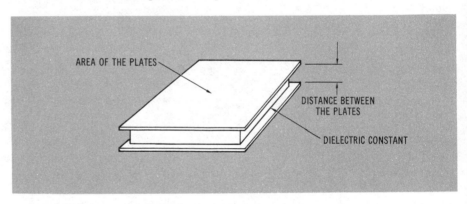

The larger the area of the plates, the greater the number of electrons required to bring the capacitor up to full charge. Also, moving the plates closer together increases the capacitance. Regarding the type of dielectric material, some materials (such as mica and Mylar) have a much higher "dielectric constant"—ability to concentrate electric fields—than other materials (such as air). Capacitors using materials having a high dielectric constant have larger capacitance values than capacitors using materials of lower dielectric constant.

Measurement

The basic unit of measurement for capacitance is the farad (F). A 1-F capacitor charges or discharges at the rate of 1 V per second when the current is fixed at 1 A. Very few practical capacitors are rated anywhere close to 1 F. Most are much smaller. For this reason, most capacitors are rated in terms of microfarads (one-millionth of a farad) or picofarads (one-millionth of a microfarad). The abbreviations for microfarad and picofarad are μF, and pF, respectively.

Voltage Ratings

If the voltage applied to a capacitor is too large, the dielectric fails to maintain its insulating qualities. It breaks down under this stress of electrostatic force and allows current to flow through the dielectric material.

Capacitors are assigned a working-voltage rating. This rating is the highest voltage that can be applied to the capacitor without risking a dielectric breakdown. The type of material used for the dielectric and the distance between the plates of the capacitor determine the working-voltage rating.

It is standard practice to use a capacitor that has a voltage rating that is at least twice the highest amount of voltage expected in the circuit. So it isn't unusual to find a capacitor with a voltage rating of 50 V used in a circuit that uses no more than 15 V.

CAPACITORS USED IN TIMING CIRCUITS

A capacitor and a resistor can be placed in a circuit to operate as a timing device. If the values of resistance (R) and capacitance (C) are carefully selected, the circuit can determine when an exact number of seconds has elapsed. *Figure 9-9* shows the basic elements of such a circuit.

This circuit contains a 15-V dc source, a 1-MΩ resistor, a double-throw switch, and a 20-μF capacitor. When the switch is set to the charge position, the battery charges the capacitor. Notice that the path for charge current includes the resistor. This means the capacitor will charge at a slower rate than it would if the resistor were replaced with a straight piece of wire.

When the switch is set to the discharge position, the capacitor is disconnected from the battery and is allowed to discharge through the resistor. The capacitor will discharge at a slower-than-normal rate because the resistor limits the amount of discharge current.

The basic unit of measurement for capacitance is the farad. Most capacitors are rated in microfarads and picofarads.

The voltage rating of a capacitor is an indication of the largest amount of voltage that can be applied to the plates without risking breakdown of the dielectric material that separates the plates.

Charging a capacitor through a resistor slows down the charge time.

Discharging a capacitor through a resistor slows down the discharge time.

**Figure 9-9.
Capacitor Timing
Circuit**

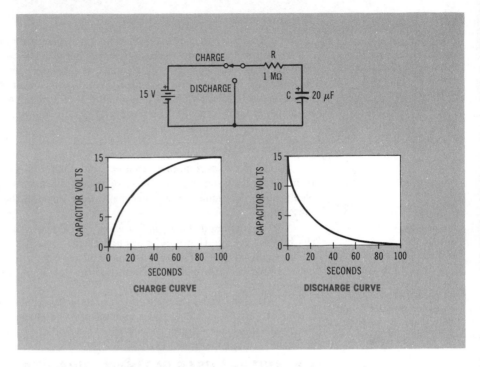

If you construct this circuit and use the voltmeter as shown across the capacitor, you can observe the voltage on the capacitor. When the capacitor first begins to charge, you will see the voltage rising rapidly at first. But as the value approaches 15 V (the full-charge value), you will clearly see the pointer on the voltmeter slowing down. Likewise, when you begin discharging the capacitor, the voltage drops rapidly at first, then slows down as it approaches 0 V. The graphs in *Figure 9-9* show the change in voltage across this capacitor as time passes. Notice that it takes about five seconds to charge the capacitor and the same amount of time to discharge it.

The amount of time it takes to charge and discharge a capacitor through a resistor is approximately 5 times the resistor value multiplied by the capacitor value. In this example, R is 1 MΩ and C is 20 μF, so the charge and discharge time is: 5 × 1 MΩ × 20 μF = 100 seconds. If you replace the capacitor with one having a value of 1 μF, the charge/discharge time becomes 5 × 1 MΩ × 1 μF = 5 seconds.

The time required to charge or discharge a capacitor through a resistor is approximately equal to 5 times the value of the resistor multiplied by the value of the capacitor.

RC Time Constant

The product of the resistance times the capacitance in such a circuit is known as the "RC time constant." The time constant of a circuit is equal to the amount of time required for changing the voltage on the capacitor by 63.2%. The capacitor charges about 63% of the full value during the first time constant. Then it charges about 63% of the remaining value during the second time constant. Then it charges 63% of the

The RC time constant of a circuit is equal to the value of the resistor multiplied by the value of the capacitor.

remaining value during the third time constant, and so on. In theory, the capacitor is never fully charged. For all practical purposes, however, the capacitor becomes fully charged after five consecutive time constants (5 times the value of the resistor multiplied by the value of the capacitor).

The capacitor discharges in the same way: it loses about 63% of its full charge during the first time constant, 63% of the remaining voltage during the second time constant, and so on.

Figure 9-10 shows the charge and discharge curves that apply to all RC timing circuits. The graph shows the percentage of charge on the capacitor as time passes. The passage of time is expressed in terms of time constants.

**Figure 9-10.
RC Time Constant**

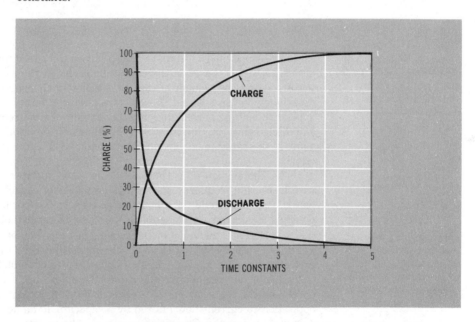

The time constant or charge and discharge times of an RC timing circuit do not depend on the amount of applied voltage.

It is especially important that you realize that the amount of voltage applied to an RC circuit has absolutely no bearing on the time constant or the charge and discharge times. One might think that increasing the amount of applied voltage would decrease the charge time for the capacitor. Increasing the amount of voltage certainly increases the amount of charge current but because the capacitor has to charge to a higher voltage, the amount of time is the same in any case.

CAPACITIVE REACTANCE

The RC timing circuits described in the previous discussion use dc voltage sources. The capacitor charges when the source voltage is applied to it, and it discharges when the plates are connected. Capacitors and resistors perform a different, but equally important, function in ac circuits. *Figure 9-11* shows a capacitor connected to an ac power source.

Figure 9-11.
Ac Capacitor Circuit

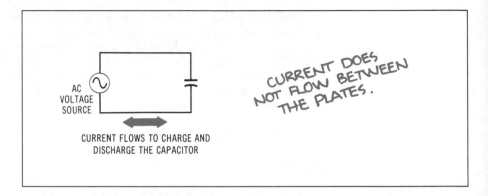

AC VOLTAGE SOURCE

CURRENT FLOWS TO CHARGE AND
DISCHARGE THE CAPACITOR

CURRENT DOES NOT FLOW BETWEEN THE PLATES.

The amount of current flowing in an ac capacitor circuit depends on the frequency of the applied voltage and the value of the capacitor.

In an ac circuit, the applied voltage changes polarity at a regular rate. This constantly changing voltage causes the capacitor to charge, discharge, and recharge in the opposite polarity at the same rate. Recalling that it takes some amount of time to completely charge and discharge a capacitor, you can see that there is a relationship between how often the applied ac voltage is changing and the amount of time required for charging and discharging the capacitor. The frequency of the applied voltage and the value of the capacitor determine the amount of ac current flowing in the circuit.

The amount of ac current flowing in a capacitor circuit increases with increasing frequency and increasing values of capacitance. In other words, you can double the amount of ac current flowing in such a circuit by doubling the frequency or doubling the value of capacitance.

The opposition to ac current flow provided by a capacitor is called "capacitive reactance." The basic unit of measurement for capacitive reactance is the ohm.

Because the value of capacitance has a powerful effect on the amount of current flowing in an ac circuit, it is possible to say that the capacitance is offering some kind of opposition to current flow. The larger the value of the capacitor or the higher the operating frequency, the less opposition the capacitor offers to current flowing through the circuit. This opposition to current flow offered by a capacitor in an ac circuit is called capacitive reactance. Capacitive reactance, like resistance, is measured in ohms.

Capacitive reactance is inversely proportional to the applied frequency and value of capacitance.

Capacitive reactance is inversely proportional to the applied frequency and value of the capacitor. That is, increasing frequency or capacitance decreases capacitive reactance, whereas decreasing frequency or capacitance increases the amount of capacitive reactance.

Impedance

Combined resistance and capacitive reactance in an ac circuit provide an overall opposition to current flow that is called impedance. Impedance is measured in units of ohms.

Capacitors are rarely used alone in an ac circuit. Rather, they are used with other components, such as a resistors. *Figure 9-12* shows a capacitor connected in series with a resistor and an ac power source. This circuit offers two different kinds of opposition to current flow: the resistance of the resistor and the capacitive reactance of the capacitor. These two kinds of opposition to current flow are called "impedance." Like resistance and capacitive reactance, impedance is measured in ohms.

**Figure 9-12.
Ac RC Circuit**

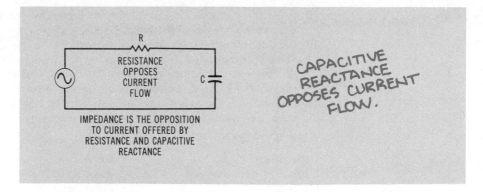

R

RESISTANCE
OPPOSES
CURRENT
FLOW

C

CAPACITIVE
REACTANCE
OPPOSES CURRENT
FLOW.

IMPEDANCE IS THE OPPOSITION
TO CURRENT OFFERED BY
RESISTANCE AND CAPACITIVE
REACTANCE

WHAT HAVE WE LEARNED?

1. A basic capacitor is made up of two conductors separated by a dielectric.
2. Current does not flow through the dielectric material of a capacitor. However current can flow onto and off of the plates.
3. When a capacitor is charging, electrons that are forced onto one plate repel electrons from the other plate. Charging continues until the source of charge current is interrupted or the voltage on the capacitor equals the source voltage.
4. A charged capacitor can hold its voltage even after the capacitor is removed from the circuit.
5. When a capacitor is discharging, the plate having the excess of electrons returns them through an external circuit to the plate having the shortage of electrons. The discharge is complete when there is no longer any difference in potential between the two plates.
6. Current flows in a capacitor circuit only when there is a change in voltage.
7. A capacitor blocks direct current but allows alternating current to charge and discharge its plates.
8. The basic unit of measurement for capacitance is the farad (F). More commonly used units are the microfarad (μF = one-millionth of a farad) and the picofarad (pF = one-millionth of a microfarad).
9. The voltage rating of a capacitor is an indication of how much voltage can be applied to the plates without risking a breakdown of the dielectric material.
10. The voltage ratings of capacitors are often selected so that the value is at least twice the largest amount of voltage found in the circuit.
11. Charging and discharging a capacitor through a resistor slows down the charge and discharge times.
12. A capacitor charges more slowly as it reaches its full-charge voltage, and it discharges more slowly as it reaches its discharge voltage.
13. The charge and discharge time for an RC circuit is approximately equal to 5 times the value of the resistor multiplied by the value of the capacitor.
14. The time constant of an RC circuit is equal to the value of the resistor multiplied by the value of the capacitor.
15. The amount of voltage has no bearing on the time constant or charge and discharge times of an RC timing circuit.

16. A capacitor opposes ac current flow as capacitive reactance. Capacitive reactance is measured in ohms.
17. Capacitive reactance is inversely proportional to the applied frequency and value of capacitance: increasing the values of frequency or capacitance decreases the capacitive reactance, whereas decreasing values of frequency or capacitance increases the amount of capacitive reactance.
18. Impedance is the overall opposition to current flow offered by resistance and capacitive reactance in an ac circuit. Impedance is measured in ohms.

KEY WORDS

Capacitance	Farad
Capacitive reactance	Impedance
Capacitor	Microfarad
Dielectric	Picofarad
Dielectric constant	RC time constant

Quiz for Chapter 9

1. A basic capacitor is made up of:
   a. two dielectrics separated by a conductor.
   b. two coils of wire separated by a dielectric.
   c. two conductors separated by a dielectric.
   d. a single coil of wire wrapped around a dielectric material.

2. Which one of the following statements best describes the process of charging a capacitor?
   a. Current flows through the external circuit until there is no longer any difference in charge between the two plates of the capacitor.
   b. Current flows from the voltage source, through the dielectric material of the capacitor, and away from the positive plate.
   c. Current flows from the plate having an excess number of electrons, through the dielectric, and to the plate having a shortage of electrons.
   d. Current flows through the external circuit until the voltage across the plates of the capacitor matches the voltage of the source.

3. Which one of the following statements best describes the process of discharging a capacitor?
   a. Current flows through the external circuit until there is no longer any difference in charge between the two plates of the capacitor.
   b. Current flows from the voltage source, through the dielectric material of the capacitor, and away from the positive plate.
   c. Current flows from the plate having an excess number of electrons, through the dielectric, and to the plate having a shortage of electrons.
   d. Current flows through the external circuit until the voltage across the plates of the capacitor matches the voltage of the source.

4. The basic unit of capacitance is the:
   a. ampere.
   b. farad.
   c. ohm.
   d. volt.

5. The working voltage of a capacitor:
   a. is the least amount of voltage you can apply in order to get the capacitor to function properly.
   b. is an indication of its capacity for storing electrons.
   c. must not exceed twice the value of the largest amount of voltage in the circuit.
   d. is the largest amount of voltage that should ever be applied to it.

6. What is the RC time constant of a circuit composed of a 1-μF capacitor and a 10-MΩ resistor?
   a. 0.1 second.
   b. 0.5 second.
   c. 1 second.
   d. 5 seconds.
   e. 10 seconds.
   f. 50 seconds.

7. How long will it take to discharge a 100-μF capacitor through a 1-MΩ resistor?
  a. 0.01 second.
  b. 0.05 second.
  c. 1 second.
  d. 5 seconds.
  e. 100 seconds.
  f. 500 seconds.

8. Doubling the voltage applied to an RC timing circuit:
  a. doubles the charge time.
  b. cuts the charge time in half.
  c. has no effect on the charge time.

9. Doubling the frequency of an ac voltage applied to a capacitor:
  a. doubles the amount of capacitive reactance.
  b. cuts the capacitive reactance in half.
  c. has no effect on the capacitive reactance.

10. Doubling the value of the capacitor in an ac circuit:
  a. doubles the amount of capacitive reactance.
  b. cuts the capacitive reactance in half.
  c. has no effect on the capacitive reactance.

11. The basic unit of capacitive reactance is the:
  a. ampere.
  b. farad.
  c. ohm.
  d. volt.

12. The basic unit of impedance is the:
  a. ampere.
  b. farad.
  c. ohm.
  d. volt.

Understanding Semiconductors

ABOUT THIS CHAPTER

Semiconductors have revolutionized the electronics industry and, consequently, our society during the past quarter century. First, small, efficient, and economical transistors replaced larger, inefficient, and expensive vacuum tubes. Then many transistor operations were replaced with even more efficient and economical integrated circuits. And now integrated circuits are becoming even smaller and more useful than ever before. When you complete this chapter, you will have a basic understanding of some of the most common of these semiconductor devices —diodes, transistors, and integrated circuits.

CREATING A SEMICONDUCTOR

A semiconductor material is neither a good conductor nor a good insulator.

You have already learned that a good conductor is a material that allows current to flow easily through it. On the other hand, a good insulator is a material that does not allow current to flow easily. This chapter introduces you to a material that is not quite a conductor, but not quite an insulator, either. This special material is called a "semiconductor" material.

All modern semiconductors are artificial. The process begins with a material that is normally considered a good insulator. The most popular material is silicon, the basic constituent of ordinary beach sand and window glass. This insulating material is melted at high temperatures and purified to a level better than 99.9%. Before the melted material is allowed to cool, it is "doped" with a minute quantity of selected impurities.

Without these impurities, the molten silicon would simply cool and form crystals of silicon—a very good electrical insulator. The added impurities, however, completely alter the electrical nature of the material. Instead of crystallizing as non-conductive silicon, this mixture crystallizes to form a semiconductor material.

N-Type and P-Type Materials

A semiconductor material is "doped" to produce an n-type or p-type material. Current is carried by electrons in the n-type material and by positively charged "holes" in the p-type material.

As far as the electrical properties are concerned, there are only two basic kinds of semiconductor materials—n-type and p-type materials. Both conduct electrical current, but do so in different ways. N-type semiconductor materials carry current in the form of electrons. P-type semiconductor materials carry current in the form of positive charges called "holes." Whether a semiconductor is of the n or p type depends on the kind of impurity that is added to the molten insulator material during the manufacturing process.

Figure 10-1 shows that n and p semiconductor materials can carry electrical current of both polarities, with the current flowing as electrons through the n-type material and as "holes" through the p-type material.

**Figure 10-1.
Charge Carriers in N
and P Semiconductor
Materials**

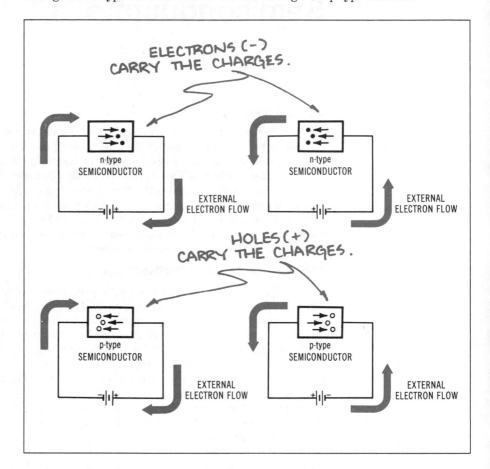

What is a semiconductor? You have just learned that it is a special kind of electrical material and that there are just two kinds of active semiconductor materials—n and p. But there is another definition for semiconductor. The term also refers to an electrical device that takes advantage of the special electrical properties of semiconductor materials. This device is called a "diode."

DIODES

A diode is one of the simplest and most useful of all semiconductor devices. As shown in *Figure 10-2*, it is made of a block of semiconductor material that has n properties at one terminal and p properties at the other. The region where the n- and p-types meet is called the "pn junction."

Figure 10-2.
Semiconductor Diode

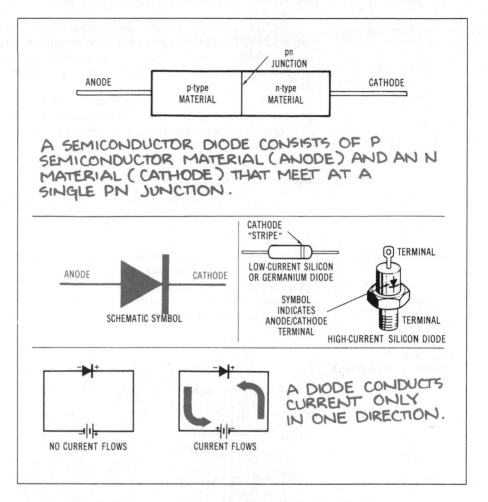

A SEMICONDUCTOR DIODE CONSISTS OF P SEMICONDUCTOR MATERIAL (ANODE) AND AN N MATERIAL (CATHODE) THAT MEET AT A SINGLE PN JUNCTION.

A DIODE CONDUCTS CURRENT ONLY IN ONE DIRECTION.

The n material in a diode is the cathode; the p material is the anode. The two materials share a common pn junction.

The terminal connected to the p-type material is called the "anode"; the terminal connected to the n-type material is the "cathode." The schematic symbol for a diode looks like a triangle pointing to a bar. The base of this triangle is the anode connection, and the bar is the cathode connection.

Low-current diodes look much like a .25-W resistor and have about the same diameter as a thick pencil lead. A single stripe painted around the plastic body indicates the cathode connection. By the simple process of elimination, the second terminal must be the anode. High-current diodes are much larger and have sturdy metal housings. A diode symbol painted on the side of these diodes indicates which terminal is the anode and which is the cathode.

Current flows in only one direction across a pn junction. This junction allows electrons to flow easily from the n to the p material (from cathode to anode), but the junction blocks electron flow from the p to the n junction (from anode to cathode).

The special feature of a diode is that it conducts current only in one direction. Recall that when used alone n and p materials conduct current in either direction. Put an n and a p material together to form a pn junction, however, and current can flow through the device in only one direction.

Electron current flows fairly easily from the n side of the pn junction to the p side—from n to p, from cathode to anode. Current cannot flow through a diode when the source voltage is connected such that the cathode is more positive than the anode. Current flows rather easily through a diode when the anode is more positive than the cathode.

Converting Ac to Dc

Diodes are used wherever it is important to allow current to flow in only one direction, as in rectifier circuits.

Rectifier Circuits

Rectifier circuits use diodes to convert ac to dc. This action relies on the fact that a diode can pass current in only one direction.

A rectifier circuit converts ac to dc. You know that an alternating voltage increases and decreases in a positive polarity, then increases and decreases in the negative polarity. *Figure 10-3* shows three complete cycles of ac voltage. The job of a rectifier circuit is to convert these alternating voltages into a smooth dc voltage of the kind available directly from batteries.

**Figure 10-3.
Cycles of an Ac Voltage**

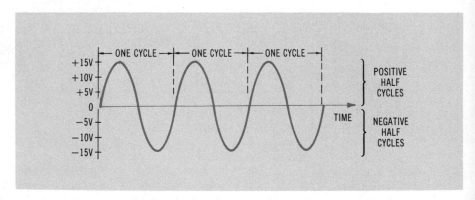

How does a diode affect the flow of alternating current? Remember that current flows in only one direction through a diode—when the anode is more positive than the cathode. *Figure 10-4* shows how a simple diode affects the flow of current. During the positive half cycle, the anode of the diode is positive with respect to its cathode. This is the condition required for current flow in a diode. As a result, current flows through the load resistor, and the ac source voltage appears across that resistor.

**Figure 10-4.
Half-Wave Rectifier
Action**

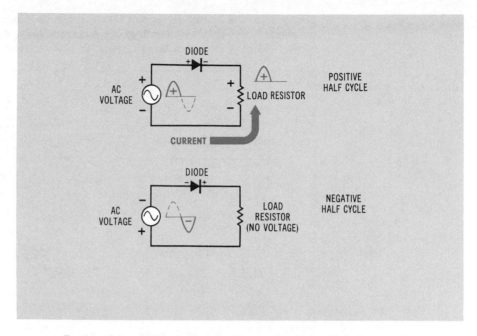

During the negative half cycle, the anode of the diode is negative with respect to the cathode. The diode cannot conduct current in that direction. So there is no current flow through the circuit, and there is no voltage across the load resistor.

Because the diode allows current to flow in only one direction, the full cycle of alternating voltage applied to this rectifier circuit appears as half-wave pulses across the load resistor. This circuit is called a "half-wave rectifier." You will learn about full-wave rectifiers in Chapter 12.

Diode Ratings

Semiconductor diodes are generally rated according to the amount of current they can handle and the amount of reverse voltage that can be applied without risking breakdown of the pn junction. The current rating applies to the amount of current a diode can handle when it is conducting electrons in the "forward" direction—from cathode to anode. The voltage rating applies to the amount of voltage that can be applied in the "reverse" direction (anode negative, cathode positive) without breaking down the pn junction and allowing the diode to conduct "against the grain."

Figure 10-5 shows that a normally conducting diode has a small voltage across it. This forward-conduction voltage is not very large, and it does not increase very much as the current increases through the diode. Bear in mind, however, that power (dissipated in the form of heat) is equal to voltage multiplied by current. Passing an excessive amount of current through a diode creates more heat than it can dissipate. As a result, there is a good chance that the diode can destroy itself. So diodes are rated according to the amount of current they can handle before they become overheated. Typical current ratings are on the order of 500 mA for small diodes and 10 A for rather large ones.

Figure 10-5.
Diode Current and
Voltage Ratings

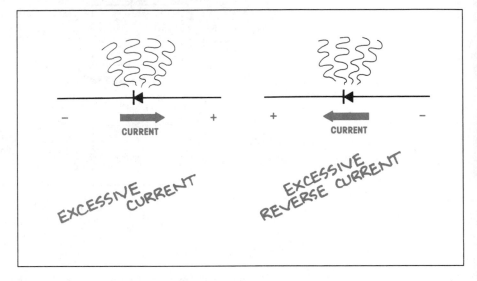

Just as any semiconductor diode can be destroyed if you allow too much current to flow through it, it can be destroyed by an excessive amount of reverse voltage. A diode is supposed to conduct electrons only from cathode to anode. You have seen that a diode in a rectifier circuit blocks the flow of current in the opposite direction. But no diode is perfect, and there is a limit to its ability to block the flow of current. This limitation is expressed in terms of its "peak inverse voltage" (PIV). PIV ratings can be as low as 50 V for some diodes and as high as 800 or 1000 V for others.

SOME OTHER TYPES OF DIODES

Other types of semiconductor diodes serve purposes that are often entirely separate from rectifying operations. *Figure 10-6* shows the schematic symbols for the diodes described here.

Light-Emitting Diodes

Charges crossing the pn junction of a light-emitting diode create visible light.

One very popular kind of special diode is called a "light-emitting diode," or an LED. Charges crossing the pn junction in an LED produce visible light—usually red. This light is not generated in the same fashion as an ordinary light bulb generates light. Recall that current flowing through the filament in a light bulb produces heat and that light results when the filament is heated to a white-hot temperature. Most of the energy consumed by a light bulb is dissipated as useless heat. An LED is a poor cousin of the laser and, as such, does not require heat in order to generate its light in an efficient fashion.

Figure 10-6.
Other Types of
Semiconductor Diodes

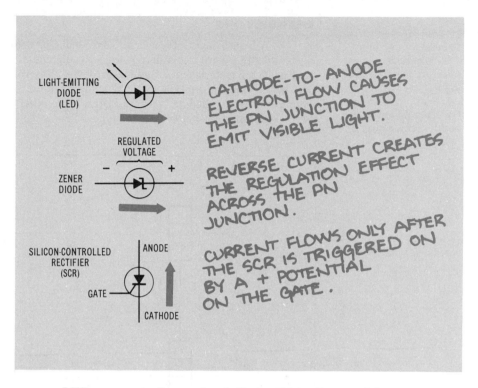

LEDs are most often used as indicator lights. Groups of tiny LEDs can be arranged in such a way that they can be lighted in patterns representing numerals and letters of the alphabet.

Zener Diodes

Zener diodes are used as voltage regulator devices. They actually conduct current in reverse—from anode to cathode.

"Zener diodes" are used as voltage regulators. Once an applied voltage reaches a specified level, it is difficult to cause any further increase in the voltage across this device. Zener diodes are available with regulation voltage ratings such as 6, 12, 24, and 48 V. These diodes are actually operated in reverse. As long as you do not exceed their zener current ratings, their normal direction for electron flow is from anode to cathode.

Silicon-Controlled Rectifiers

Silicon-controlled rectifiers do not conduct current from cathode to anode until a small trigger current is applied to a gate connection. Once gated on, an SCR conducts until the cathode-to-anode voltage is removed or reversed.

A "silicon-controlled rectifier" (SCR) is a diode that does not conduct in either direction until a small current is applied to a gate connection. Once an SCR is "gated on" it conducts current from cathode to anode until the voltage applied between the cathode and anode is removed or reversed. SCRs are most often used as high-current electronic switches.

TRANSISTORS

Transistors are three-layer semiconductor devices. An npn transistor has a p layer between two n layers, and a pnp transistor has an n layer between two p layers.

A transistor is a three-layer semiconductor device. As shown in *Figure 10-7* there are two different ways to combine n- and p-type semiconductor materials in a three-layer device. The result is two kinds of transistors. One has a p-type material sandwiched between two n-type materials. This is called an "npn transistor." The other kind fits a section of n-type semiconductor between two p-types. This is called a "pnp transistor."

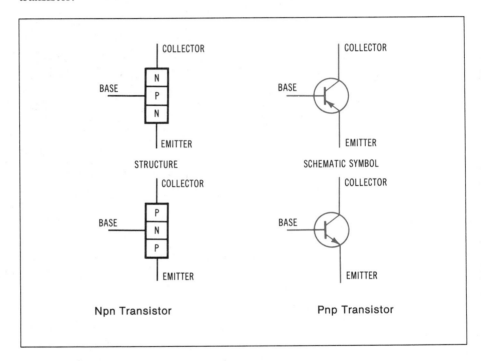

Transistors provide you with electrical connections to all three layers. The connections to the two outer layers are called the "emitter" and "collector." The connection to the middle layer is called the "base." Notice in *Figure 10-7* that the only difference between the schematic symbols for npn and pnp transistors is the direction of the arrow at the emitter terminal.

The electrical connections to the three layers of semiconductor materials in a transistor are called the emitter, base, and collector.

As long as there is no voltage applied to the base of a transistor, one of the two pn junctions blocks current flow between the emitter and collector.

Recall that a semiconductor diode has just one pn junction and that electrons can flow through the diode only when the n side of the junction is negative compared with the p side of the junction. *Figure 10-8* shows that a transistor has two different pn junctions—an emitter-base junction and a collector-base junction. As long as there is no voltage applied to the base, current cannot flow between the emitter and collector. No matter which polarity of voltage you apply between the emitter and base connections, one of these two junctions prevents current from flowing through the device.

**Figure 10-8.
Two Junctions in a
Transistor**

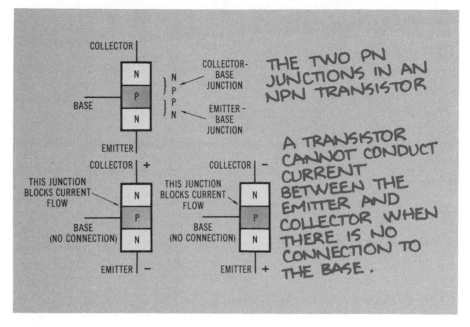

Biasing

Figure 10-9 shows the important role the base plays in the operation of a transistor. Whenever the base is made positive with respect to the emitter, the emitter-base pn junction allows current to flow between those two terminals. This is called a "forward-biased junction." But when you reverse the polarity, the pn junction prevents current from flowing through the emitter-base circuit. This is called a "reverse-biased junction." Generally speaking, the emitter-base circuit in a transistor works like a diode.

Forward-biasing the emitter-base junction in a transistor allows current to flow between the emitter and collector.

Now suppose that you forward-bias the emitter-base junction of an npn transistor. This requires applying a voltage to the base that is positive with respect to the emitter. Further suppose that you apply a positive voltage to the collector as well. Under these conditions something important takes place. Some of the electrons flowing through the emitter-base junction fill the holes in the p material and effectively eliminate the collector-base junction. With this junction electrically eliminated, as shown in *Figure 10-10*, the transistor also conducts current from the emitter to the collector. In short, forward-biasing the emitter-base junction of a transistor allows current to flow between the emitter and collector.

**Figure 10-9.
Biasing the Emitter-
Base Junction**

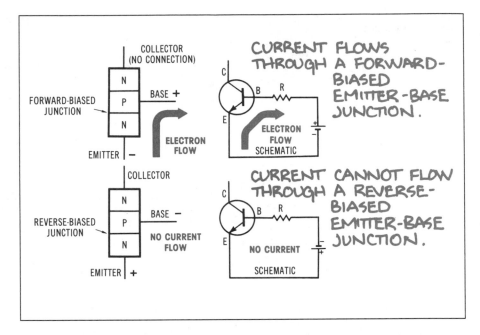

**Figure 10-10.
Currents in a
Conducting Npn
Transistor**

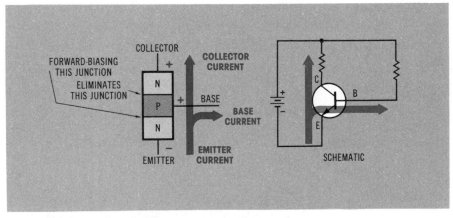

The same principles apply to pnp transistors. The only difference is the direction of current flow through the emitter-base junction and between the emitter and collector.

Notice in the illustration that the emitter supplies current to both the base and emitter connections. Emitter current is the sum of the base current and collector current. Transistors are designed so that the collector current is much greater than the base current—often as much as 500 times greater. This means you can control the amount of collector current with a relatively small amount of base current. A change of 5 μA of current in the

A small change in base current can cause a relatively large change in collector current.

emitter-base circuit, for example, can allow the collector current to change by as much as 2500 μA, or 2.5 A. This is the basis for the use of transistors in amplifier circuits.

Transistors as Amplifiers

One of the most common applications of transistors is in amplifier circuits. *Figure 10-11* illustrates the action of two different transistor amplifier circuits. One uses an npn transistor and the other a pnp transistor.

**Figure 10-11.
Transistor Amplifier
Circuits**

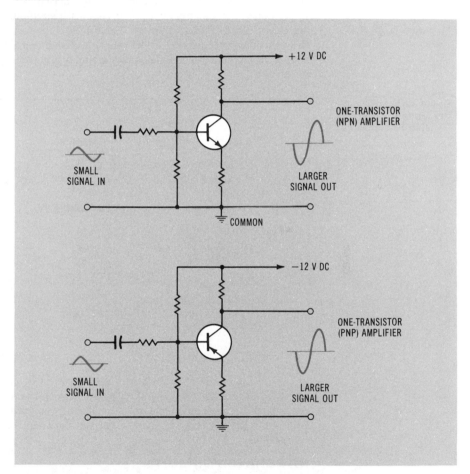

The output of one transistor amplifier can be connected to the input of the next in order to multiply the amplifying effect. It is not unusual to see simple amplifiers built from eight or ten individual transistors. Transistors are smaller, more efficient, and less expensive than vacuum tubes. For these reasons, it has been possible to improve the quality and lower the cost of electronic circuits by using transistors in place of vacuum tubes.

Alternative Types of Transistors

The transistors described thus far in this chapter are called "bipolar junction transistors," or simply bipolar transistors. Bipolar transistors are mainly current-operated devices—the amount of base current determines the amount of collector current. Another kind of semiconductor transistor controls current with an applied voltage. These are called "field-effect transistors," or FETs. A more recent refinement of the field-effect transistor is a "metal-oxide-silicon transistor," or MOSFET. *Figure 10-12* compares the schematic symbols and common names for the terminals.

Figure 10-12.
The Modern Family of Transistor Devices

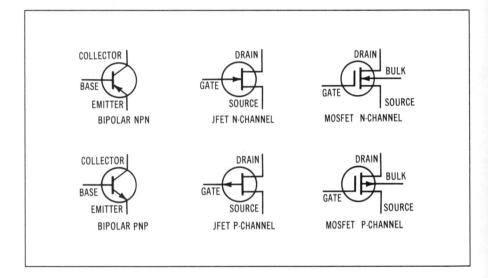

FET and MOSFET devices use electrical fields at a gate connection to control the flow of charges between the source and drain connections. A very small amount of charge at the gate connection can control a relatively large amount of current between the source and drain connections.

INTEGRATED CIRCUITS

You have already learned that it is possible to produce a variety of useful semiconductor devices by placing one or two pn or MOS junctions on a small piece (or chip) of silicon. But why stop with just a couple of these junctions? Why not go for dozens, hundreds, or even thousands of separate junctions and use the conductive properties of doped silicon to interconnect all of these junctions to create entire circuits on a single chip? That is what an "integrated circuit" is.

The heart of an integrated circuit is a tiny chip of silicon that has many separate junctions and resistances etched onto it. These elements are interconnected with etched conductors to form entire electronic circuits.

The heart of an integrated circuit (IC) is a tiny chip of silicon that has had multiple layers of bipolar or field-effect junctions etched onto it. Lines of conductive silicon are used as conductors for interconnecting the junctions, and sections of less conductive silicon are used as resistors. Gold or gold-plated beam leads are welded to the points on the chip that are to be connected elsewhere in the outside world. This assembly is enclosed in a tough epoxy case, and sturdy pins are provided for making the electrical connections and holding the IC device firmly in a socket assembly.

Some of the simpler IC devices contain the equivalent of two dozen transistors, diodes, and resistors. Some of the more complex "chips," such as those used in microcomputers, have the equivalent of thousands of separate devices.

Figure 10-13 shows a typical eight-pin IC package. The connection pins are numbered in a standard sequence. A mark in the epoxy case indicates pin number 1. Some ICs have as few as four pins and as many as forty. Specifications sheets for each IC device indicate the purpose of each numbered pin.

Figure 10-13.
Typical IC Package

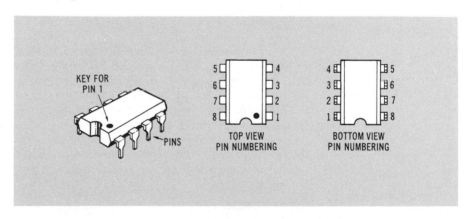

You should be aware that IC technology is almost single-handedly responsible for the recent revolution in electronics. Thousands of power-gobbling vacuum tubes or tens of thousands of connections to separate transistors can be reduced to a single IC package that costs just a few dollars.

WHAT HAVE WE LEARNED?

1. A semiconductor material is neither a good conductor nor a good insulator.
2. Electrons carry the electrical charges in n-type semiconductor materials; holes carry the charges in p-type materials.
3. A semiconductor diode consists of n and p materials that meet at a common pn junction. The n material is called the cathode and the p material the anode.

4. Electron current flows only from the n to the p side of a pn junction. This means a diode conducts easily when the cathode is negative and the anode is positive. It does not conduct current when the polarity of the applied voltage is reversed.
5. Rectifier circuits use diodes to convert ac to pulsating dc.
6. Semiconductor diodes are rated according to the maximum amount of forward-conducting current (in milliamperes and amperes) and the maximum amount of reverse voltage, called the peak inverse voltage, they can withstand (in volts).
7. Light-emitting diodes (LEDs) generate visible light when charges cross their pn junctions.
8. Zener diodes regulate a voltage that is applied in reverse across them.
9. Silicon-controlled rectifiers (SCRs) do not conduct until they are "gated on" by a small current applied to a gate connection. Once triggered into conduction, an SCR can be turned off only by removing or reversing the polarity of the voltage applied between the cathode and anode.
10. Transistors are three-layer semiconductors that are classified as npn and pnp transistors.
11. Electrical connections to the three layers of a transistor are called the emitter, base, and collector.
12. A transistor can conduct current between the emitter and collector only as long as the emitter-base junction is forward biased.
13. A small change in the base current of a transistor can cause a relatively large change in the amount of collector current.
14. Field-effect transistors (FETs) and metal-oxide-silicon transistors (MOSFETs) use a voltage charge on a gate terminal to control the flow of charges between the source and drain connections.
15. An integrated circuit (IC) has many bipolar or field-effect junctions, resistors, and conductors that form complete electronic circuits on a single chip of silicon.

KEY WORDS

Anode	Metal-oxide-silicon transistor (MOSFET)
Base	Npn transistor
Bipolar transistor	Peak inverse voltage (PIV)
Cathode	Pn junction
Collector	Pnp transistor
Emitter	Rectifier circuit
Field-effect transistor (FET)	Reverse bias
Forward bias	Semiconductor
Integrated circuit (IC)	Silicon-controlled rectifier (SCR)
Light-emitting diode (LED)	Zener diode

Quiz for Chapter 10

1. Which one of the following statements best describes the difference between n- and p-type semiconductor materials?
   a. N-type materials are good semiconductors, whereas p-type materials have holes that make them good insulators.
   b. N-type materials are made of silicon; p-type semiconductors are always made of germanium.
   c. Charges are carried by electrons in n-type materials and by holes in p- type materials.
   d. Charges are carried by holes in n-type materials and by electrons in p- type materials.

2. In a semiconductor diode:
   a. the terminal connected to the n material is the cathode, and the terminal connected to the p material is the anode.
   b. the terminal connected to the p material is the cathode, and the terminal connected to the n material is the anode.
   c. the terminal connected to the p material is the base.
   d. the terminal connected to the n material is the base.

3. In a diode, electrons flow more easily from:
   a. the p material to the n material.
   b. from the cathode to the base.
   c. from the anode to the cathode.
   d. from the cathode to the anode.
   e. from the base to the anode.

4. Rectifier circuits use diodes to:
   a. convert ac to dc.
   b. convert dc to ac.
   c. increase the voltage level of a circuit.
   d. increase the current level of a circuit.
   e. increase the power level of an ac signal.

5. Which one of the following devices emits visible light when current flows from cathode to anode?
   a. FET.
   b. LED.
   c. SCR.
   d. Zener diode.

6. Which one of the following devices is commonly used as a voltage regulator?
   a. FET.
   b. LED.
   c. SCR.
   d. Zener diode.

7. Which one of the following devices has to be triggered into conduction?
   a. FET.
   b. LED.
   c. SCR.
   d. Zener diode.

8. In an npn, transistor:
   a. the emitter represents the p material, and the base and collector represent the n materials.
   b. the base represents the p material, and the emitter and collector represent the n materials.
   c. the collector represents the p material, and the emitter and base represent the n materials.

9. In a pnp transistor:
  a. the emitter represents the n material, and the base and collector represent the p materials.
  b. the base represents the n material, and the emitter and collector represent the p materials.
  c. the collector represents the n material, and the emitter and base represent the p materials.

10. In a MOSFET transistor:
  a. a charge applied to the source connection controls the current flowing between the gate and drain connections.
  b. current applied to the emitter-base junction controls the current flowing between the emitter and collector connections.
  c. a charge applied to the gate connection controls the current flowing between the source and drain connections.
  d. a charge applied to the drain connection controls the current flowing between the gate and source connections.

Understanding Cathode-Ray Tubes

ABOUT THIS CHAPTER

In this chapter you will learn about the principles of operation of a cathode-ray tube, or CRT. It is important that you learn these principles because CRTs are the "picture tubes" in television receivers and the "screens" for modern computers and radar systems. When you complete your work in this chapter, you will understand how CRTs generate and control a fine beam of electrons to produce images on the screen. You will also appreciate the fact that CRTs are the most commonly used survivors of the age of vacuum-tube technology.

VACUUM TUBES

Before the advent of practical semiconductors (transistors and integrated circuits), electronics technology was built around vacuum tubes. Vacuum tubes once performed the tasks of modern-day transistors and diodes. The superior performance and economy of semiconductors have rendered vacuum-tube technology practically obsolete. One important vacuum-tube device remains, however—the cathode-ray tube (CRT).

CRTs convert electrical signals into detailed visual information.

CRTs are used wherever it is necessary to translate electrical signals into detailed visual information. As shown in *Figure 11-1*, CRTs are used as the screens for television receivers, computers, oscilloscopes, and radar.

CRTs work on the same basic principles as vacuum tubes.

Because the operation of modern CRTs is based on vacuum-tube technology, you should have some understanding of basics of vacuum tubes before beginning your study of how CRTs work and how they are used today.

Electron Current Flow in a Vacuum

Electron current can be made to flow through a vacuum.

Air is a poor conductor of electron current, but a vacuum is a fairly good conductor. Given the right circumstances, you can force electrons to flow through a vacuum. These circumstances are illustrated for you in *Figure 11-2*.

**Figure 11-1.
Some Applications of
CRTs**

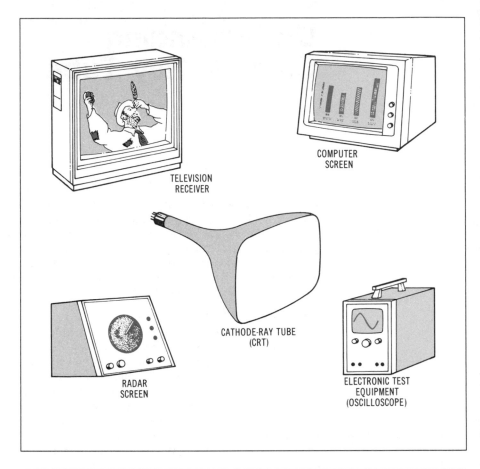

**Figure 11-2.
A Primitive Vacuum
Tube**

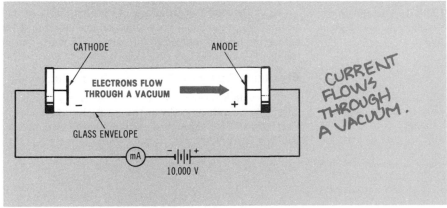

The tube, itself, consists of a glass envelope that has most of the air drawn out of it. This satisfies the need for a vacuum. Electrodes, called the "cathode" and "anode," are inserted into the vacuum through air-tight seals. Connecting these electrodes to a high-voltage power supply provides the electrical force required for making electrons flow through the tube. Provided the vacuum is good enough and the applied voltage is high enough, the milliammeter will register a constant dc current flow through this circuit.

Electron current flows from the cathode to the anode in a vacuum-tube device.

Electron current in a vacuum tube flows from the cathode to the anode. This is because the cathode is connected to the negative terminal of the voltage source, and the anode is connected to the positive source. In any kind of vacuum tube, electron current flows from the cathode to the anode.

The only way to increase the amount of current flowing through this primitive vacuum tube is by increasing the amount of vacuum or increasing the amount of voltage applied to the electrodes. Neither of these procedures is very practical. *Figure 11-3* shows a more practical way to force more current to flow through the vacuum.

**Figure 11-3.
An Improved Vacuum
Tube Uses a Heater**

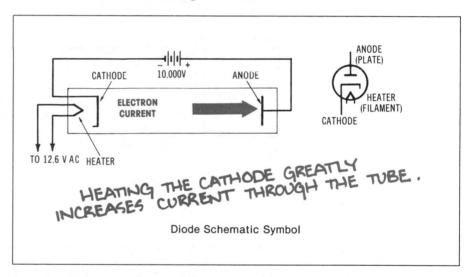

Diode Schematic Symbol

Heating the cathode in a vacuum tube increases the amount of current flowing through the tube.

Recall that raising the temperature of a conductor "excites" its electrons, thereby increasing the number of free electrons. Heating the cathode in a vacuum-tube device greatly increases the amount of current that can flow from the cathode to the anode. Cathodes are also coated with a special material that further enhances the ease with which the cathode produces free electrons.

The heater is normally connected to a power source that is entirely separate from the high-voltage power supply. The heater's power source is usually 6.3 V or 12.6 V taken from the secondary of an ordinary filament transformer.

Controlling Electron Flow in a Vacuum Tube

The basic elements of a vacuum tube, as described thus far, include an anode, a cathode, a heater, and a glass envelope to secure those elements in a vacuum. The external components of this system are the high-voltage dc power supply and a low-voltage ac power source for operating the heater. These components are adequate for causing a useful amount of current to flow through the tube, but there are no provisions for controlling that current in a precise fashion.

Controlling the amount of current flow through a vacuum tube requires the addition of another element—a "control grid." As shown in *Figure 11-4*, the control grid is inserted into the path for current flow through the vacuum. The control grid is not a solid piece of metal, but rather a screen or spiral of wire. The grid material, itself, does not impede the flow of electrons.

**Figure 11-4.
A Vacuum Tube with a
Control Grid.**

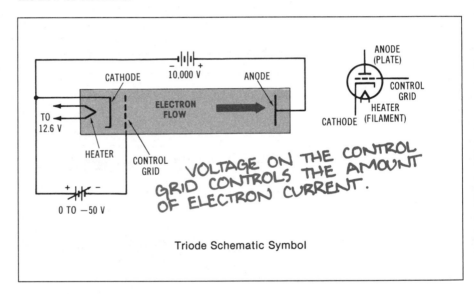

Triode Schematic Symbol

A negative voltage on the control grid slows the rate of electron flow from cathode to anode. The larger the negative voltage on the control grid, the smaller the amount of current through the tube.

Notice that the control grid is connected to a variable dc power source. The power source is connected so that the control grid is negative compared with the cathode. Recalling that like charges repel one another, you can see that the negative voltage on the control grid reduces current flow through the vacuum tube by repelling the electrons. The larger the amount of negative charge on the control grid, the lower the amount of current flowing from the cathode to the anode. If the negative voltage on the control grid is large enough, it is possible to block the tube's current flow altogether.

VACUUM-TUBE PRINCIPLES APPLIED TO A CRT

The basic principles just described for simple vacuum tubes apply directly to the operation of a CRT. *Figure 11-5* illustrates a simple CRT and the necessary external circuitry. The only additional part of the scheme is the phosphor screen—the round or rectangular "face" of the CRT.

**Figure 11-5.
A Primitive CRT**

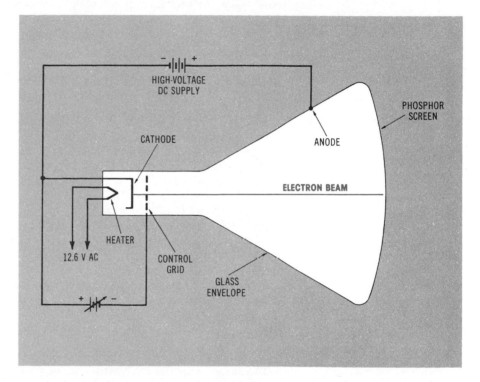

The heater, cathode, and control grid are not essentially different from those described for simple vacuum tubes. The heater warms the cathode, the cathode provides the source of electrons for the beam, and the voltage on the control grid determines the amount of electron current in the beam.

The anode in a CRT is not a simple piece of metal. Rather it is a thin metallic coating that is "flashed" onto the inner surface of the cone-shaped part of the glass envelope. A button extending through the glass provides the necessary electrical connection to the high-voltage power supply. The high positive voltage on the anode attracts electrons from the cathode at a high rate of speed. These accelerated electrons do not stop until they strike the phosphor on the screen.

The phosphor on the screen of a CRT converts the beam of electrons into a spot of visible light. The larger the amount of beam current, the brighter the spot appears. The voltage on the control grid determines the brightness of the spot of light on the screen of a CRT. Increasing the amount of negative voltage on the control grid decreases the brightness of the spot.

The color of the spot of light on the face of a CRT is determined by the kind of phosphor that is used. A monochrome CRT generates varying brightnesses of just one color.

A color (multicolor) CRT uses different phosphors for red, green, and blue colors.

The electron gun in a CRT is an assembly that contains all the components for generating the electron beam and controlling its brightness and quality.

The focusing anode in an electron gun makes it possible to apply a voltage that compresses the electron beam to produce a sharper spot of light on the screen.

The phosphor on the screen of a CRT is a material that emits light when electrons strike it. It is painted onto the inside surface of the face of the screen. This phosphor glows with a prescribed color wherever the electron beam strikes it. The narrower the electron beam, the smaller the point of light appears on the screen. The larger the amount of current in the electron beam, the brighter the spot of light appears.

Recall that you can adjust the amount of current through a vacuum tube by adjusting the amount of negative voltage applied to the control grid. The larger the amount of control voltage, the smaller the amount of current in the beam. This principle applies directly to the operation of a CRT, but we can state it in terms of the brightness of the spot on the face of the screen. Increasing the amount of negative voltage on the control grid reduces the amount of beam current; reducing the amount of beam current reduces the brightness of the light on the screen.

The color of the spot of light on the face of a CRT depends on the type of phosphor used. Some phosphors generate white light, some generate green light, and others generate blue, red, or yellow light. A given phosphor can generate only one color. A CRT that can produce just one color is called a "monochrome CRT." It makes no difference what the color is. A black-and-white television receiver, for example, uses a monochrome CRT having a phosphor that creates white light. Many computer screens use a monochrome CRT that produces green light.

How do color (or multicolor) CRTs work? Color CRTs, such as those used in color television sets, actually use three different phosphors: one each for red, green, and blue. These three colors can be combined in certain proportions to produce the effect of any visible color. The proportions are adjusted by controlling the voltages applied to three separate control grids and electron beams.

ELECTRON GUNS

The heater, cathode, and control grid for a CRT are all included in an assembly called the "electron gun." However there are more elements in the electron guns for modern CRTs than we have described so far. Generally speaking, the electron gun is an assembly that includes the elements necessary for creating the electron beam and controlling its intensity and quality.

For example, electron guns also include a "focusing anode." The focusing anode is located near the control grid and surrounds the electron beam. Voltages applied to the focusing anode narrow the beam to produce a sharper spot of light on the screen.

Another electrode located in the electron gun is called the "acceleration anode." A high positive voltage on this anode increases the velocity of the electrons before leaving the gun. (Once the electrons leave the gun, they are further accelerated by the main anode.)

So the electron gun in a CRT can be a fairly complicated device. The most complex electron guns are found in full-color CRTs. Full-color CRTs use three separate electron beams, and each beam has its own control grid, focusing anode, and acceleration anode.

ELECTRON-BEAM MOVEMENT

The CRT device described thus far is capable of producing a sharply defined spot of light in the middle of the screen. You can vary the brightness of the spot by varying the amount of voltage applied between the cathode and control grid. That stationary point of light is not very useful, however. The beam and the spot of light it creates on the screen must be made to move in order to convey useful information to the viewer.

There are two different mechanisms available for moving the electron beam to any desired point on the screen. You can deflect an electron beam by applying a voltage that attracts or repels the electron beam, or you can apply an external magnetic field that also deflects the beam.

Electrostatic Beam Deflection

Figure 11-6 shows a set of four electrodes that are added to a CRT that uses the principles of "electrostatic beam deflection." External voltages applied to the deflection plates attract or repel the negatively charged electron beam from its normal path through the center of the tube. Voltages applied to the horizontal deflection plates move the electron beam in the horizontal (back-and-forth) direction; voltage applied to the vertical deflection plates move the beam in the vertical (up-and-down) direction.

**Figure 11-6.
An Electrostatic-
Deflection CRT**

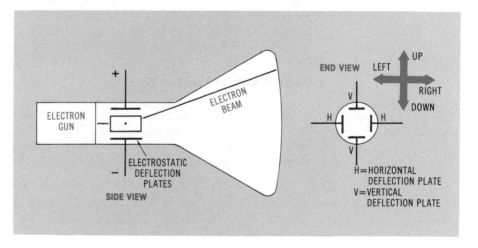

Voltages applied to the horizontal deflection plates in a CRT move the beam left or right of center. The direction of motion depends on the polarity of the applied voltage. An ac voltage of sufficient frequency creates the impression of a straight horizontal line.

Suppose that you apply no voltage to the vertical deflection plates but +100 V to the left horizontal plate and −100 V to the right horizontal deflection plate. How does this situation affect the electron beam? Bearing in mind that the electron beam is made up of electrons—negatively charged particles—you can see that the beam will be attracted to the positive voltage on the left horizontal plate and repelled by the negative voltage on the right plate. The overall effect is that the beam moves to the left. Reverse the polarity on these plates, and the beam will move to the right. Apply an ac voltage (one that switches polarity at a regular rate), and the beam will move back and forth. If the frequency of the ac voltage is about 15 Hz or more, the moving beam will appear as a straight horizontal line across the screen.

Voltages applied to the vertical deflection plates move the beam up or down, depending on the polarity of that applied voltage.

In a similar fashion, applying voltages to the vertical deflection plates position the electron beam in the vertical direction. A positive polarity on the upper deflection plate attracts the beam upward; a positive polarity on the lower deflection plate attracts the beam downward. Applying an ac voltage to the vertical deflection plates can create the visual impression of a straight vertical line.

The amount of deflection depends on the amount of voltage applied to the deflection plates. The larger the amount of voltage, the larger the amount of deflection.

So the polarity of the voltages applied to the horizontal and vertical deflection plates determines the direction the electron beam moves. It is equally important to understand that the amount of voltage determines the amount of deflection. The greater the amount of voltage applied to the deflection plates, the farther the electron beam moves.

In an electrostatic-deflection CRT, the position of the beam is determined by the amount and polarity of the voltages applied to the deflection plates. A proper selection of these voltages makes it possible to position the beam anywhere on the screen. Add the ability to control the brightness of the spot by varying the voltage applied to the control grid, and you have full control over the beam and its appearance on the face of the CRT.

Magnetic Beam Deflection

An electron beam can be deflected by external magnetic fields.

Just as an electron beam can be deflected by the forces of electrostatic attraction and repulsion, it can be deflected by magnetic fields. This is the principle behind the operation of CRTs that use magnetic deflection. *Figure 11-7* shows a CRT that uses magnetic deflection. Notice the electromagnet coils wrapped around the neck of the tube.

**Figure 11-7.
A Magnetic-Deflection
CRT**

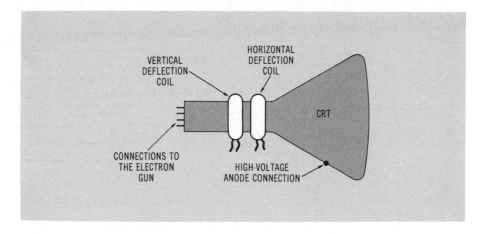

The horizontal deflection
coil deflects the electron
beam in the horizontal direction, whereas the ver-
tical deflection coil
deflects the beam in the
vertical direction. The
amount of current deter-
mines the amount of
deflection.

There are two magnetic "deflection coils"—one for deflecting the beam horizontally and one for deflecting it vertically. Current flowing through the horizontal deflection coil positions the electron beam in the horizontal direction. Whether this coil deflects the beam to the right or left depends on the direction of current flow through the coil. The vertical deflection coil positions the electron beam in the vertical direction, with the direction of current flow determining whether the beam is deflected upward or downward. The amount of current flowing through the deflection coils determines the amount of deflection.

Most CRTs in use today use magnetic deflection. The tubes are simpler because the deflection coils are added to the outside. This also makes them less expensive. There are some other special advantages of magnetic deflection that need not be discussed here. Some types of special electronic testing equipment use electrostatic deflection, but modern electronics heavily favors the use of magnetic deflection.

THE CREATION OF IMAGES

You have already learned that the electron beam in a CRT produces a single spot of light where it strikes the phosphor. Creating complex images or filling a computer screen with letters and numerals is a matter of positioning the beam and controlling its brightness at a very rapid rate. This action takes place so rapidly that your eyes and brain are fooled into seeing the entire screen as patterns of light and dark (or color).

Raster Scan Pattern

One of the most commonly used patterns of motion for the electron beam is called the "raster scan" pattern. This pattern is built into the operation of all television receivers and most computer screens. *Figure 11-8* illustrates this beam-deflection pattern.

**Figure 11-8.
The Raster Scan
Pattern**

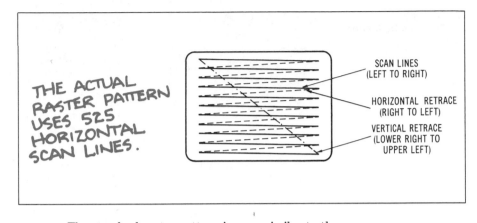

THE ACTUAL RASTER PATTERN USES 525 HORIZONTAL SCAN LINES.

SCAN LINES
(LEFT TO RIGHT)

HORIZONTAL RETRACE
(RIGHT TO LEFT)

VERTICAL RETRACE
(LOWER RIGHT TO
UPPER LEFT)

The standard raster pattern is very similar to the one your eyes use when reading a printed page such as this one. Your eyes begin in the upper-left corner of the page and run left to right along the first line. When you reach the end of the first line on the page, your eyes snap down and to the left—to the beginning of the next line. You continue that pattern until you reach the end of the last line on the page.

The raster scan pattern sweeps the beam of a CRT in a systematic left-to-right, top-to-bottom pattern.

The raster pattern begins near the top of the screen and progresses in a line-by-line, left-to-right fashion until the beam reaches the bottom of the screen. The standard television and computer display systems have 525 scan lines. The system scans all 525 lines on the "page" 30 times a second.

As the electron beam is moving from left to right across the screen, changes in the amount of beam varies the brightness of the light. However the beam is always turned off during the time it is being brought back to the starting point for the next line. This process of returning from the end of one line to the beginning of the next is called the "retrace." A horizontal retrace takes place at the end of each left-to-right scan, and a vertical retrace takes place when the beam reaches the bottom of the screen and snaps back to the top to begin another frame.

Blanking Pulses

Blanking pulses turn off the electron beam during the horizontal and vertical retrace intervals.

We do not want the beam to be visible during the horizontal and vertical retrace intervals, so the beam is turned off during that time. The pulses that are used to turn off the electron beam during the retrace intervals are called "blanking pulses."

Synchronizing Pulses

Circuits that generate the horizontal and vertical sweeping waveforms for the deflection coils are built into the system that includes the CRT. The operation of these waveform generators are kept in step, or synchronized, by means of "synchronizing pulses." There are synchronizing pulses for both the horizontal and vertical sweep waveform generators.

So a signal that is fed to a CRT system consists of video information that is to be plotted as a picture in the screen, blanking pulses to turn off the beam during the retrace intervals, and synchronizing (or "sync") pulses to keep the beam sweeping circuits in proper step. Video signals for color CRTs also include the color information.

Composite Video Signals

Such signals that are applied to the antenna connections of a television receiver are called "composite video" signals. Composite video signals usually include audio information as well. Television signals are always broadcast or sent through commercial TV cable systems as composite video signals. Many personal computers have a composite video output jack that lets you view the computer information and generate the sounds through a standard television receiver.

WHAT HAVE WE LEARNED?

1. Cathode-ray tubes (CRTs) convert electrical signals into detailed visual information.
2. CRTs work on the same basic principles as do vacuum tubes.
3. Electron current can be made to flow through a vacuum.
4. Electron current flows from the cathode to the anode in a vacuum-tube device.
5. Electron current through a vacuum-tube device can be greatly increased by heating the cathode.
6. Applying a negative voltage on the control grid reduces current flow through a vacuum tube by repelling some of the electrons leaving the cathode. The larger the negative voltage on the control grid, the smaller the amount of cathode-to-anode current.
7. The phosphor painted on the inside surface of the face of a CRT converts the energy of speeding electrons into visible light.
8. The size of the spot of light on the face of a CRT is determined by the diameter of the electron beam. The narrower the electron beam, the smaller the spot.
9. The brightness of the spot of light on the face of a CRT is determined by the amount of current in the electron beam. The greater the amount of beam current, the brighter the spot.
10. A given phosphor can generate light of only one color. A monochrome CRT uses a single kind of phosphor.
11. Full-color CRTs use three different phosphors: one for red, one for green, and one for blue.
12. Voltages applied to the focusing anode, located in the electron gun, compress the electron beam and produce a smaller, more clearly defined spot of light on the screen.
13. The electron beam in a CRT can be moved by means of voltage or magnetic deflection.
15. A CRT that uses electrostatic (voltage) deflection has a set of horizontal and vertical deflection plates built into it.

16. The electron beam in an electrostatic-deflection CRT is attracted by positive voltages on the deflection plates and repelled by negative voltages on the plates.
17. Voltages applied to the horizontal deflection plates in an electrostatic-deflection CRT move the electron beam in the horizontal direction. Voltages applied to the vertical deflection plates move the beam in the vertical direction.
18. In an electrostatic-deflection CRT, the amount of deflection is determined by the amount of voltage applied to the deflection plates. The larger the amount of voltage, the greater the amount of deflection.
19. A magnetic-deflection CRT uses the magnetic fields from external electromagnets to position the beam.
20. The vertical deflection coil positions the beam in the vertical direction, whereas the horizontal deflection coil positions the beam in the horizontal direction.
21. The direction of current flow in the horizontal deflection coil determines whether the beam is deflected to the right or left. The direction of current flow in the vertical deflection coil determines whether the beam is deflected up or down.
22. The amount of current flowing in the deflection coils determines the amount of deflection. The greater the amount of current, the greater the amount of deflection.
23. Most CRTs in use today use magnetic deflection.
24. The standard raster scan pattern sweeps the beam in a CRT through a systematic left-to-right, top-to-bottom pattern.

KEY WORDS

Acceleration anode	Electrostatic deflection
Anode	Filament
Blanking pulses	Focusing anode
Cathode	Heater
Cathode-ray tube (CRT)	Horizontal deflection
Color CRT	Monochrome CRT
Composite video	Phosphor
Control grid	Raster scan
Deflection coils	Sync pulses
Deflection plates	Synchronizing pulses
Diode	Vertical deflection
Electron gun	

Quiz for Chapter 11

1. A CRT is:
   a. an electromechanical device.
   b. a vacuum-tube device.
   c. a semiconductor device.
   d. a nuclear device.

2. In any vacuum-tube device, electron current flows:
   a. from the positive electrode to the negative electrode.
   b. from the anode to the cathode.
   c. from the cathode to the anode.
   d. along the outer surface of the glass envelope.

3. The purpose of the heater in a vacuum tube is to:
   a. prevent the buildup of moisture on the cathode.
   b. "boil" additional electrons off the cathode.
   c. decrease the air pressure inside the glass envelope.
   d. ensure a good airtight seal.

4. Which one of the following statements most accurately describes the function of the control grid in a vacuum tube?
   a. A negative voltage on the control grid reduces the amount of current flow between the cathode and anode by repelling electrons in the stream.
   b. A positive voltage on the control grid reduces the amount of current flow between the cathode and anode by attracting electrons away from the stream.
   c. A negative voltage on the control grid increases the amount of current flow between the cathode and anode by injecting more electrons into the stream.

5. The control grid in a CRT determines the:
   a. brightness of the spot of light on the screen.
   b. size of the spot of light on the screen.
   c. color of the spot of light on the screen.
   d. position of the spot of light on the screen.

6. Which one of the following components of a CRT is *not* included in the electron gun?
   a. heater.
   b. cathode.
   c. control grid.
   d. focusing anode.
   e. anode.

7. Where can you expect to see the spot of light located on the face of an electrostatic-deflection CRT when the upper vertical plate is maximum positive and the right horizontal plate is maximum negative?
   a. Center of the screen.
   b. Top of the screen.
   c. Bottom of the screen.
   d. Upper-left corner of the screen.
   e. Upper-right corner of the screen.
   f. Lower-left corner of the screen.
   g. Lower-right corner of the screen.

8. What determines the amount of beam deflection in an electrostatic-deflection CRT?
   a. The frequency of the voltage applied to the deflection plates.
   b. The amount of voltage applied to the control grid.
   c. The amount of voltage applied to the deflection plates.
   d. The polarity of the voltage applied to the deflection plates.
   e. The frequency of the voltage applied to the control grid.

9. What determines whether the beam will be deflected to the right or left in a magnetic-deflection CRT?
   a. The amount of current flowing through the horizontal deflection coil.
   b. The amount of current flowing through the vertical deflection coil.
   c. The direction of current flowing through the horizontal deflection coil.
   d. The direction of current flowing through the vertical deflection coil.

10. What determines the amount of deflection of the electron beam in a magnetic-deflection CRT?
    a. The amount of current flowing in the deflection coils.
    b. The direction of current flowing in the deflection coils.
    c. The frequency of current flowing in the deflection coils.
    d. The direction of voltage applied to the deflection plates.
    e. The amount of voltage applied to the deflection plates.

Basic Circuit Actions

ABOUT THIS CHAPTER

When studying electronics, it is important that you pause occasionally to take stock of what you have learned and that you try to pull it together in some meaningful way. This is also important in that it prepares you for learning new material. This chapter uses many of the fundamental ideas you have already learned. You will see how they apply to various kinds of circuits, and you will learn some new principles that are important for completing your work in this book.

POWER SUPPLIES

Most of the basic components within electronic equipment require a dc power source. This requirement poses no problem for battery-operated equipment because batteries naturally supply dc voltage and current. However, it is not always possible nor desirable to operate equipment directly from batteries. In such instances it is necessary to call upon the most commonly used source of electrical energy—the local power utility company.

Using power from your local electric company (through the plug on the wall) eliminates the need for using and maintaining batteries. The only problem is that your power company supplies ac electricity. It is thus necessary to convert that readily available ac power to dc power that is required for most electronic components. For this reason, most kinds of electronic equipment have a built-in, rectifier power supply.

Electronic power supplies convert available levels of ac power to a desired level of dc power. These power supplies vary a great deal in cost and complexity, depending on the necessary level of quality and reliability.

Ac power is readily available but must be converted to dc power as required for most components in a piece of electronic equipment.

Half-Wave Rectifier Power Supply

The simplest kinds of power supplies use a single diode as a rectifier. This "half-wave rectifier" circuit converts full-wave ac into half-wave dc. As shown in *Figure 12-1*, a capacitor connected across the dc circuit goes a long way toward smoothing out the "humps" in the rectified waveform. The bit of ac that remains on the output of this power supply is called "ac ripple." It is almost impossible to eliminate all the ripple in the dc power from such a simple circuit.

**Figure 12-1.
Half-Wave Rectifier
Power Supply**

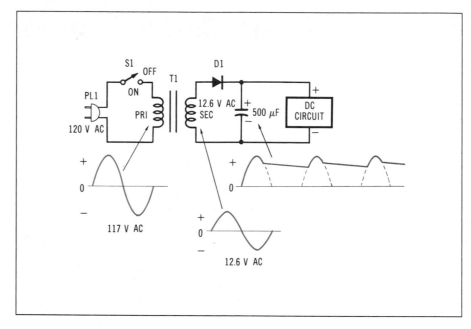

Full-Wave Rectifier Power Supply

Whereas a half-wave rectifier circuit uses only one of the half cycles of an ac waveform, a "full-wave rectifier" circuit manages to use both half cycles. As indicated in *Figure 12-2*, a full-wave power supply constructed from diodes is more complex that the half-wave version. The trade-off is that it is easier to lower the level of ac ripple from the output of a full-wave rectifier circuit.

The relative complexity of full-wave rectifier circuits that are built from two or four diodes is misleading, however. Today's component technology provides a full-wave rectifier circuit in a single package. This package, sometimes called a full-wave bridge rectifier, greatly simplifies the construction and maintenance procedures for power supplies.

**Figure 12-2.
Full-Wave Rectifier
Power Supplies**

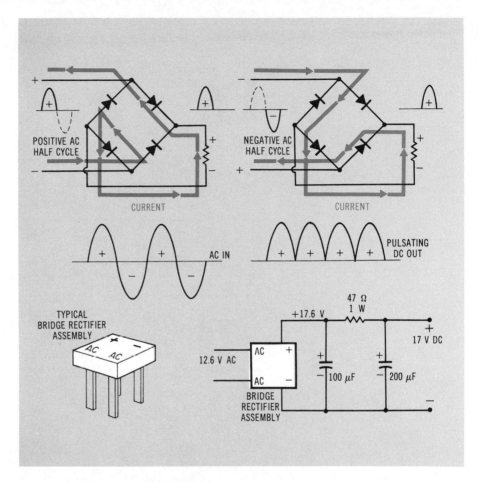

Regulated Power Supplies

Regulated power supplies, properly used, can eliminate all ac ripple from the rectifier circuit and maintain a fixed dc output voltage in spite of normal fluctuations in the ac utility power. Voltage regulators are not widely available in packages that make them look like three-terminal power transistors. These regulators have specified output voltages such as 5 V, such as that shown in *Figure 12-3*, 12 V, and 24 V. As long as the rectified voltage applied to the input terminals is always greater that the specified output voltage, the input voltage can fluctuate a great deal (including ac ripple) without affecting the regulated output voltage level.

**Figure 12-3.
Regulated Power
Supply**

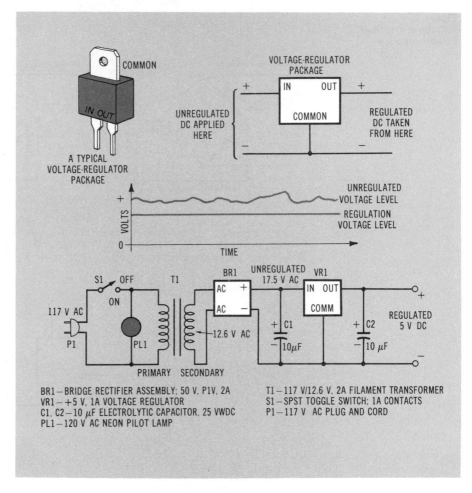

Bridge rectifier and voltage regulator devices are so economical these days that regulated power supplies are frequently more economical than non-regulated power supplies that use simple diodes and require large and expensive filter capacitors.

AMPLIFIERS

An amplifier is any circuit that boosts the voltage or current level of a signal. Chapter 10 shows how transistors are capable of amplifying small electrical signals into larger ones. Practical amplifier circuits use more than one transistor, however. Getting a desired amount of amplification without distorting the signal requires the use of more than a single transistor. In fact, most modern amplifier circuits are built around integrated circuits that each contain the equivalent of dozens of transistors.

Some Applications of Amplifiers

Perhaps one of the most familiar applications of amplifier circuits is in audio equipment. Tape players, public-address systems, and radio and television receivers all include loudspeakers that transform electrical signals into sound. Amplifiers are necessary in all these instances because the original signal is far too weak to operate a loudspeaker.

Audio Amplifiers

An ideal audio amplifier works equally well across the range of frequencies required for audio work—from 5 Hz to 25,000 Hz.

Amplifier circuits that are specifically designed for boosting the power level of audio signals are called "audio amplifiers." The best audio amplifiers work in the frequency range of 5 Hz to 25,000 Hz. Audio amplifiers of lesser quality work in a more limited band of frequencies within that range.

Audio amplifiers are rated according to the amount of power (in watts) they can deliver to the loudspeakers, the range of frequencies they can reproduce at the given output power level, and the amount of distortion and noise the components themselves introduce into the signal.

Radio-Frequency Amplifiers

Radio-frequency amplifiers work in the upper kilohertz, megahertz, and gigahertz range.

Another popular kind of amplifier circuit is designed to handle frequencies much higher than encountered in audio systems. These higher-frequency amplifiers, usually called "radio-frequency amplifiers," boost the current and voltage levels of signals commonly encountered in radio, television, and radar equipment. Some operate in the upper kilohertz and megahertz range allocated for radio and television communications. Others operated in the gigahertz range (1000 MHz). This range of frequencies is used mainly for radar, military, and satellite communications.

Radio-frequency, or rf, amplifiers are rated according to their voltage and current gain (amount of amplification), the range of frequencies they can handle, and their input and output impedances.

Other Uses

Amplifiers are also used in industrial control and robotic equipment. The internal workings of such equipment usually operate at low voltage and current levels. The devices to be controlled, however, often require much higher levels of control power. This amplification operation often begins with integrated-circuit amplifiers, goes to higher-power transistor amplifiers, and ends with special high-power semiconductor control devices such as silicon-controlled rectifiers (SCRs) and triacs. These amplifier systems are designed to amplify pulse waveforms and respond to changes in dc voltage levels.

Types of Amplifiers

The total amplification of two or more amplifier stages is equal to the product of the individual amplification factors.

Practical amplifier circuits are composed of one or more amplifier stages. Single amplifier stages are rarely designed to boost a signal more than 10 or 20 times. So where it is necessary to amplify a signal 20 times or more, you will find the output of one amplifier feeding its signal to another. The overall amplification of this multiple-stage amplifier is equal to the product of the individual amplification ratings.

Suppose, as illustrated in *Figure 12-4*, you have a circuit composed of three amplifier stages. The first has a gain of 2, the second a gain of 8, and the third a gain of 15. What is the overall amplification of the circuit? It is 2 × 8 × 15, or 240. If it is a voltage amplifier, applying a 10-mV signal at the input stage should produce a 2400-mV (2.4-V) version at the output of the final stage.

**Figure 12-4.
Three-Stage Amplifier
Circuit**

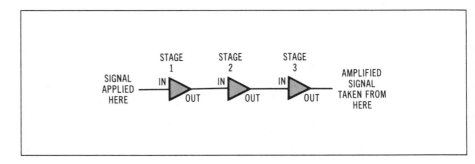

Ac amplifiers boost alternating-current signals and block any dc portion. Dc amplifiers can amplify both ac and dc signal levels.

There are many ways to classify amplifier circuits. For instance, there are ac and dc amplifiers. Ac amplifiers boost the alternating-current portion of a signal and leave the dc portion unchanged (or remove it altogether). A dc amplifier can boost the voltage or current level of dc signals.

You can usually identify an ac amplifier system by noting that one amplifier stage is "coupled" to the next by the use of capacitors or transformers as shown in *Figure 12-5*. (Recall that capacitors and transformers can pass only ac signals.) On the other hand, you can identify dc amplifiers by the fact that the output of one stage is coupled directly to the input of the next stage through either a resistor or a straight piece of conductor.

Operational Amplifiers

Operational amplifiers can be adapted to a wide range of amplifier applications.

The most common kind of IC amplifier circuit is called an "operational amplifier," or "op-amp." This amplifier can be adapted to a wide range of different amplifier circuits, depending on how you connect external components to the device.

Internally, IC operational amplifiers are high-quality dc amplifiers. Pin connections provide the means for connecting the amplifier to external components and other amplifiers. IC operational amplifiers can be used as ac amplifiers when the stages are interconnected with capacitors or transformers. *Figure 12-6* shows a typical IC operational amplifier package.

**Figure 12-5.
Ac and Dc Coupling
Between Amplifier
Stages**

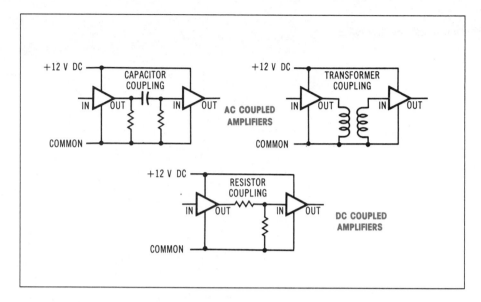

**Figure 12-6.
IC Operational Amplifier
Package**

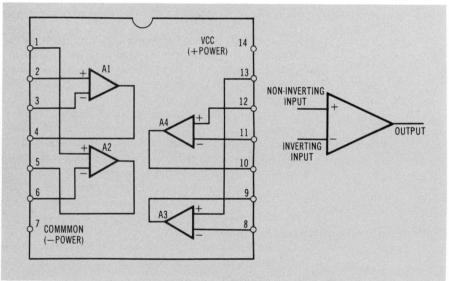

WAVEFORMS AND OSCILLATOR CIRCUITS

Types of Waveforms

Figure 12-7 shows a variety of waveforms that can perform useful functions in electronic circuitry. The "sinusoidal waveform" is the same as the one readily available from the plug on the wall. This waveform is also found in many different kinds of electronic equipment, particularly in communications equipment.

**Figure 12-7.
Common Waveforms**

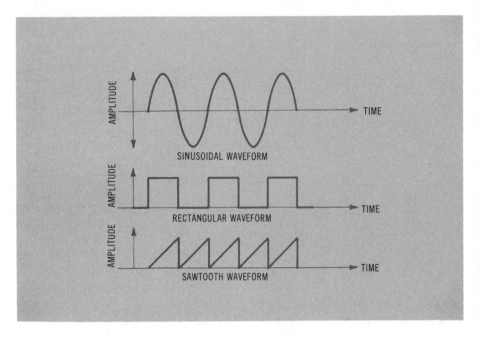

Another useful kind of waveform is the rectangular, or square, waveform. A "rectangular" waveform shows a sudden rise to a peak level, some time spent at that peak level, then a sudden drop to the base level again. These waveforms are used for timing, triggering, and electronic logic operations.

You have seen in Chapter 11 that a CRT uses a "sawtooth waveform" to move the electron beam across the face of the tube. This waveform shows a gradual increase toward a peak level followed by a sudden drop to the baseline level. Sawtooth waveforms are used for a variety of timing and control operations.

These are some of the most basic kinds of waveforms used in electronics today. There are many other kinds of waveforms, but most are variations and combinations of those shown here. It is important to notice that these waveforms have two features in common. First, they show changes in voltage or current level; second, the changes in level take place over a definite period of time.

Waveform Characteristics

Baseline and Peak

All waveforms have baseline and peak levels. A certain amount of time is required for the waveform to go through one complete cycle.

The level closest to zero is called the baseline level; those levels farthest away from zero are called the peak levels. Waveforms sometimes vary between negative and positive levels, but more often they vary between zero and some positive peak or between zero and a negative peak. A certain amount of time is required to complete one full cycle of level changes.

Period and Frequency

The time required to complete one full cycle of a waveform is called its period. The period of a waveform is measured in seconds. The period is equal to 1/frequency, and the frequency is equal to 1/period.

The time required for completing one full cycle is called the period of the waveform. The period of a waveform is measured in seconds. If the waveform is going through its pattern of level changes at a regular rate, it is also possible to specify the frequency of the waveform. The frequency of a waveform is the number of cycles completed in one second. Frequency is specified in hertz, and it is equal to 1 divided by the period. Likewise the period of a waveform can be found by dividing 1 by the frequency.

Suppose you find that the period of a sawtooth waveform is 0.01 second. What is its frequency? Well, the frequency can be found by dividing that period into 1: $\frac{1}{0.01} = 100$, or 100 Hz. If the frequency of a waveform is given as 2000 Hz, you can determine its period by $\frac{1}{2000}$, or 0.0005 second.

Oscillator Circuits

Oscillator circuits produce prescribed waveforms at the desired frequency and amplitude.

The waveforms required for electronic applications must be generated in some fashion within the equipment. Circuits that generate waveforms are called "oscillator circuits." There are about as many different kinds of oscillator circuits as there are kinds of waveforms. In any case, an oscillator circuit generates a prescribed waveform at the desired frequency and levels, or amplitudes.

LOGIC CIRCUITS

Binary Logic

A binary logic circuit is always in one of two different states: off or on.

The entire world of digital computer technology is based on the principles of "binary logic." And binary logic is based on the simple notion that a circuit has just two different states: off and on. Processing units, input/output (I/O) devices, and computer memories all work according to digital, binary principles.

Figure 12-8 illustrates the fundamental idea of binary logic. When the switch is open, current cannot flow through the series circuit, and the lamp cannot possibly light. Closing the switch completes the path for current flow, however, and the lamp lights. The circuit is either off or on. There are no "in-between" states.

Computer scientists have developed a shorthand procedure for describing the state of logic circuits. The switch, for example, has two operating conditions: open or closed. We can use the standard shorthand notation to designate an open-switch condition with a 0 and a closed-switch condition with a 1. Likewise, we can designate the lamp-off condition with a 0 and a lamp-on condition with a 1. So when the switch is in the 0 (open) state, you can see that the lamp is also in its 0 (off) state. And when the switch is in its 1 (closed) state, the lamp in this circuit is in its 1 (on) state.

Figure 12-8.
The Most Elementary
Binary Circuit

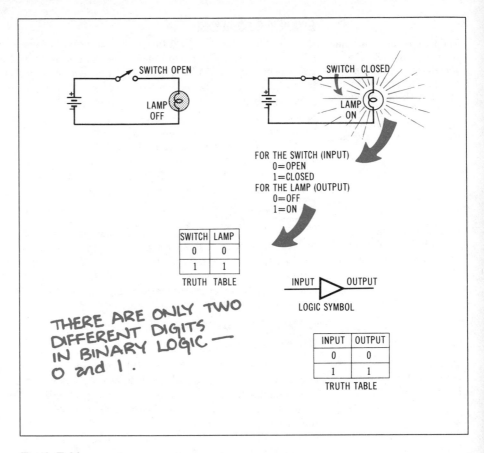

FOR THE SWITCH (INPUT)
0=OPEN
1=CLOSED
FOR THE LAMP (OUTPUT)
0=OFF
1=ON

SWITCH	LAMP
0	0
1	1

TRUTH TABLE

INPUT ▷ OUTPUT

LOGIC SYMBOL

INPUT	OUTPUT
0	0
1	1

TRUTH TABLE

THERE ARE ONLY TWO DIFFERENT DIGITS IN BINARY LOGIC — 0 and 1.

Truth Tables

A truth table completely describes the nature of a binary logic circuit in terms of the 0-and-1 logic notation.

Using this shorthand notation makes it possible to simplify a complete description of the logic circuit. The action of a logic circuit can be described completely in terms of a "truth table." As long as you understand that 0 means something (the switch or lamp) is not energized and that 1 means something (the switch or lamp) is energized, the truth table completely describes the action of the circuit: when the switch is open, the lamp is off; when the switch is closed, the lamp is on.

Although we often use switch-and-lamp circuits to introduce the fundamentals of binary logic, most circuits use transistorized switches rather than mechanical switches. *Figure 12-9* is a transistorized version of the logic operation performed by the switch in the previous example.

Whenever the voltage applied to the input (point A) is 0 V, the transistor is turned off and no current flows through resistor R. And as long as there is no current flowing through that resistor, the voltage at the output (point B) is also 0 V. When you apply +5 V to the input, however, the transistor goes into full conduction. Current flows through resistor R, and you will find nearly +5 V at the output.

**Figure 12-9.
A Transistorized Off/On
Logic Circuit**

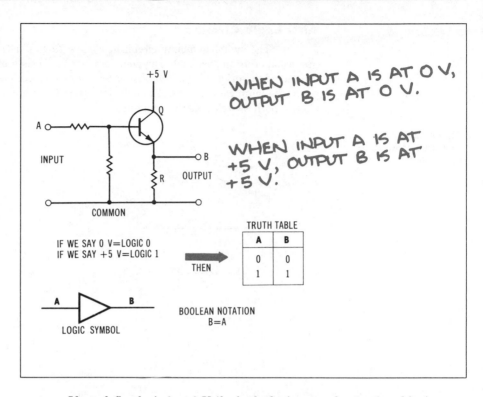

WHEN INPUT A IS AT 0 V,
OUTPUT B IS AT 0 V.

WHEN INPUT A IS AT
+5 V, OUTPUT B IS AT
+5 V.

+5 V

Q

A

INPUT

R

B

OUTPUT

COMMON

IF WE SAY 0 V=LOGIC 0
IF WE SAY +5 V=LOGIC 1

THEN

TRUTH TABLE

A	B
0	0
1	1

A   B

LOGIC SYMBOL

BOOLEAN NOTATION
B=A

Logic circuits that have the same truth table perform the same logical operation, even though the electronic principles might be vastly different.

If we define logic 0 as 0 V (for both the input and output) and logic 1 as +5 V (for both the input and output), we get the truth table shown in the diagram. Now notice that the pattern of 0s and 1s for this truth table is identical to the pattern in the previous diagram. As far as binary logic is concerned, the lamp-and-switch circuit in *Figure 12-8* is identical in function to the transistor circuit in *Figure 12-9*.

Logic Symbols

Digital engineers and technicians use standard logic symbols to simplify the drawings of logic circuits.

Just as electronics engineers and technicians use schematic symbols to simplify the drawings of electronic circuits, digital engineers and technicians use "logic symbols" to simplify their drawings. The logic symbol for the circuits we have just described looks like a triangle. The input part of the circuit is a line going to the base of this triangle, and the output is a line extending from the apex of the triangle.

Boolean Notation

The operation of a logic circuit can also be expressed in a form of a mathematical shorthand called Boolean notation.

People working in digital technology also use a form of mathematical notation called "Boolean notation" to express the operation of logic circuits. The Boolean expression for the logic circuit described in the two preceding illustrations is simply B = A. This literally says: The 0-or-1 status of output B is always equal to the 0-or-1 status of input A.

AND Logic Circuits

The switch-and-lamp circuit in *Figure 12-10* illustrates the operation of an important binary logic circuit. It consists of two switches connected in series with a lamp. As long as one (or both) of the switches is open, the lamp cannot light. The only way to get the lamp to light is by closing both switches at the same time.

**Figure 12-10.
Features of an AND
Logic Circuit**

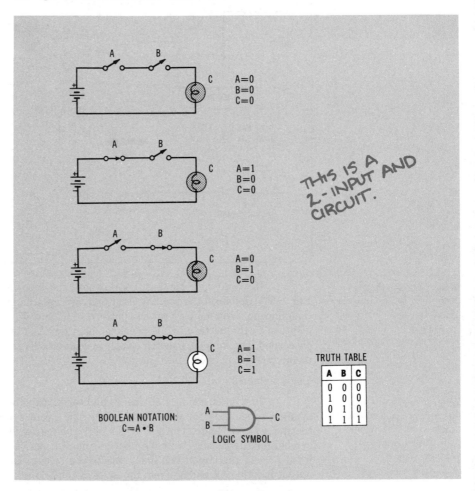

The truth table summarizes the action of this circuit. The only way to get the output (C) to go to logic 1 is by setting both inputs (A and B) to logic 1 at the same time. Any other combination of inputs shows the output at logic 0. This is the definition of a logical "AND circuit." In this case, it is a 2-input AND circuit. The output of an AND circuit is "true" only when all inputs are "true."

The logic symbol for a 2-input AND circuit shows the two inputs going to the flat side and the output leaving the rounded side. The Boolean expression uses a dot to indicate the AND operation. This expression literally says: Output C is equal to inputs A AND B.

OR Logic Circuits

If two switches connected in series with a lamp represent a useful kind of logic circuit (an AND logic circuit), perhaps connecting two switches in parallel produces another kind of logic circuit. Indeed it does, as shown in *Figure 12-11*. Notice that the lamp turns on when either (or both) of the switches is closed. The only way to get the lamp to turn off is by opening both switches at the same time.

Figure 12-11. Features of an OR Logic Circuit

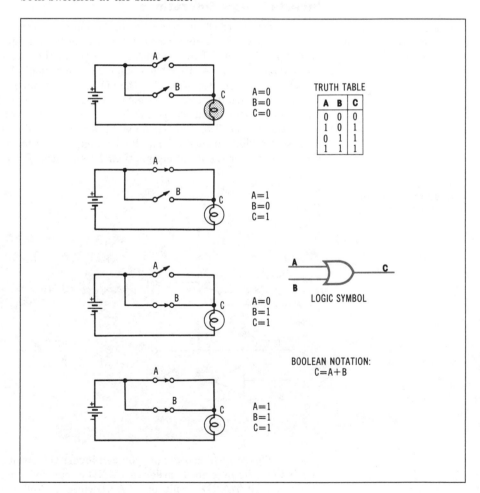

A	B	C
0	0	0
1	0	1
0	1	1
1	1	1

TRUTH TABLE

A=0
B=0
C=0

A=1
B=0
C=1

A=0
B=1
C=1

A=1
B=1
C=1

LOGIC SYMBOL

BOOLEAN NOTATION:
C=A+B

The truth table summarizes the action: Output (C) is at logic 1 when input A, input B or both are at logic 1. The output is at logic 0 only when both inputs are at logic 0 at the same time. This is the definition of a logical "OR circuit." In this case, it is a 2-input OR circuit. The output is "true" when either (or both) inputs are "true." Or to say the same thing another way, the output is "false" only when both inputs are "false" at the same time.

The logic symbol for a 2-input OR circuit shows the two inputs going to the curved side and the output leaving the pointed side. The Boolean expression uses a plus sign (+) to indicate the OR operation. This expression literally says: Output C is equal to inputs A OR B.

Inverted Logic Operations

A logic inverter circuit turns a 0 input to a 1 output and a 1 input to a 0 output.

Figure 12-12 illustrates the operation of another very important kind of logic circuit. Referring to the transistor circuit, applying 0 V (logic 0) to input A turns off the transistor. As a result the voltage at output B is +5 V (logic 1). On the other hand, applying +5 V (logic 1) to input A switches on the transistor so that the voltage at output B is very close to 0 V (logic 0). When input A is at logic 0, output B is at logic 1. And when input A is at logic 1, output B is at logic 0. Notice the pattern of 0s and 1s in the truth table. This truth table and the accompanying logic symbol represent a "logic inverter." The Boolean expression for a logic inverter shows a bar over the input designation. This is read: B equals NOT A.

**Figure 12-12.
Operation of a Logic
Inverter Circuit**

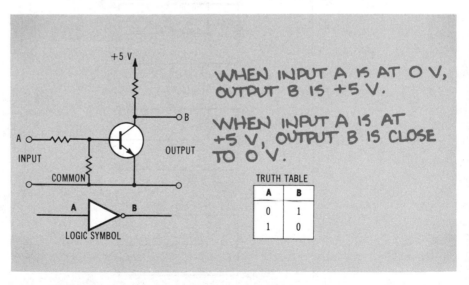

Figure 12-13 shows how you can invert the outputs of the AND and OR circuits described earlier. An AND circuit followed by an inverter produces a NOT-AND circuit, or "NAND circuit." Following an OR circuit with logic inverter produces the action of a NOT-OR circuit, or "NOR circuit."

**Figure 12-13.
Examples of NAND and
NOR Circuits**

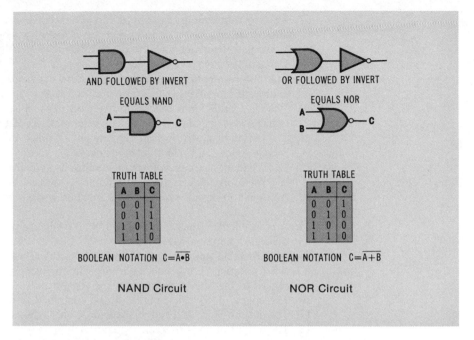

Practical Logic Devices

Logic circuits are readily available in IC packages. Most of these IC packages contain several logic devices of the same kind; some contain elaborate, commonly used combinations of logic functions. These logic ICs fall into a broad category of devices called digital ICs.

Digital IC devices that are built according to a bipolar transistor technology are called transistor-transistor logic (TTL) devices. Those using a form of FET/MOS technology are called complementary MOS, or CMOS, devices.

WHAT HAVE WE LEARNED?

1. Power supplies convert conventional ac power to dc power required for operating most kinds of electronic devices.
2. A half-wave power supply uses only one-half of the full ac waveform. A full-wave power supply uses both half-cycles of the ac waveform.
3. A regulated power supply provides a dc voltage that does not change with normal fluctuations in the ac power lines.
4. Audio amplifiers work within a range of frequencies that are important for audio work—between 5 Hz and 25,000 Hz.
5. Radio-frequency (rf) amplifiers are designed to operate in the upper kHz, MHz, and GHz (gigahertz) ranges.
6. The total amplification of two or more amplifier stages is equal to the product of the individual amounts of amplification.

7. Ac amplifiers amplify ac signals and generally ignore dc levels. You can identify ac amplifiers by the presence of capacitors or transformers in the signal path from one stage to another.
8. Dc amplifiers amplify both ac and dc signals.
9. Operational amplifiers, usually in the form of an integrated circuit, are high-quality dc amplifiers that can be adapted to a wide range of applications.
10. The most basic kinds of waveforms are sinusoidal, rectangular, and sawtooth. Other waveforms are variations or combinations of these three.
11. The period of a waveform is the amount of time, in seconds, required to complete one full cycle of level changes.
12. The frequency of a waveform is equal to 1 divided by the period; the period of a waveform can be found by dividing 1 by the frequency.
13. An oscillator circuit produces a prescribed waveform of a specified frequency and amplitude.
14. A binary logic circuit is either switched completely on or completely off. There are no in-between states.
15. A truth table completely describes the status of a logic circuit in terms of 0-and-1 notation. Truth tables portray the operation of a logic circuit without regard to the kind of technology the circuit uses.
16. The operation of logic circuits can be described in terms of Boolean notation.
17. The output of an AND circuit is logic 1 only when all inputs are at logic 1.
18. The output of an OR circuit is logic 1 when one or more inputs are at logic 1.
19. An invert logic circuit switches logic levels—0 to 1, and 1 to 0.

KEY WORDS

Ac amplifier
Amplifier
AND logic
Audio amplifier
Baseline level
Binary logic
Boolean notation
CMOS (complementary MOS) devices
Dc amplifier
Digital ICs
Full-wave rectifier
Logic inverter
Logic symbols
NAND circuit
NOR circuit

Operational amplifier
OR logic
Oscillator
Peak level
Period
Power supply
Radio-frequency (rf) amplifier
Rectangular waveform
Regulated power supply
Ripple voltage
Sawtooth waveform
Sinusoidal waveform
Truth table
TTL (transistor-transistor logic) devices

Quiz for Chapter 12

1. The main purpose of a power supply is to:
   a. boost the level of ac signals.
   b. generate sinusoidal waveforms.
   c. convert dc power to ac power.
   d. convert ac power to dc power.

2. A full-wave bridge rectifier:
   a. uses every other half cycle of an ac waveform.
   b. uses every cycle of an ac waveform.
   c. passes low frequencies around a high-frequency amplifier.
   d. passes high frequencies around a low-frequency amplifier.

3. A regulated power supply:
   a. provides a fixed level of ac voltage.
   b. maintains a fixed dc output voltage level in spite of normal fluctuations in the ac input level.
   c. increases a voltage level to a specified point.
   d. converts American 60-Hz power to European 50-Hz power.

4. The primary purpose of an amplifier is to:
   a. boost the voltage or current level of a signal.
   b. convert ac power into dc power.
   c. convert dc power into ac power.
   d. generate waveforms that serve a practical purpose within a circuit.

5. One gigahertz is equal to:
   a. 1000 Hz
   b. 10,000 Hz
   c. 1000 MHz
   d. 10,000 MHz

6. What is the overall amplification of a three-stage amplifier where each stage amplifies a signal ten times?
   a. 10.
   b. 30.
   c. 100.
   d. 300.
   e. 1000.

7. What is the period of a 20-Hz rectangular waveform?
   a. 0.01 second.
   b. 0.05 second.
   c. 2 seconds.
   d. 5 seconds.
   e. 20 seconds.

8. What is the frequency of a sawtooth waveform that has a period of 1 second?
   a. 0.01 Hz.
   b. 0.1 Hz.
   c. 1 Hz.
   d. 10 Hz.
   e. 100 Hz.

9. The output of an AND circuit is logic 1 when:
   a. none of the inputs is at logic 1.
   b. one input is at logic 1.
   c. at least half the inputs are at logic 1.
   d. more than half the inputs are at logic 1.
   e. all the inputs are at logic 1.

10. The output of an OR circuit is at logic 0 when:
    a. none of the inputs is at logic 1.
    b. one input is at logic 1.
    c. at least half the inputs are at logic 1.
    d. more than half the inputs are at logic 1.
    e. all the inputs are at logic 1.

Understanding Radio Transmitters and Receivers

ABOUT THIS CHAPTER

When you complete your study of this chapter, you will have learned what the electromagnetic frequency spectrum is, how a radio transmitter develops a broadcast signal, and how radio signals are transmitted from one place to another. You will also have learned how a radio receiver picks up transmitted signals and converts them into sound. In addition, you will have become acquainted with the differences between AM and FM radio systems.

ELECTROMAGNETIC RADIATION

Certain kinds of energy sources radiate energy. The sun, for instance, radiates energy in the form of heat, light, and cosmic rays. A radioactive material, such as uranium or plutonium, radiates energy in the form of gamma rays. An X-ray machine radiates X rays, an atom smasher radiates alpha and beta rays, and a common light bulb radiates energy in the form of heat and light. Radio transmitters also radiate energy.

These are just a few examples of the kinds of energy sources that radiate energy. Some radiate energy just a few inches. Others radiate energy across the known universe. Most of these sources are of natural origin, but a few are synthetic. We are particularly interested in radio energy—a type of synthetic radiation that scientists and engineers have learned to generate and control.

Electromagnetic waves carry radiation from place to place.

All of these forms of energy are carried from one place to another in the form of "electromagnetic waves." Electromagnetic waves carry their particular forms of energy from place to place at a speed of 186,000 miles per second.

The Electromagnetic Frequency Spectrum

The various kinds of electromagnetic radiation differ in terms of frequency.

The various kinds of electromagnetic radiation (sunlight, X rays, radio waves, and so on) differ from each other in terms of their frequencies. Recall that frequency is a measure of the number of vibrations per second. Frequency is expressed in hertz, abbreviated Hz.

Figure 13-1 shows the frequencies for various kinds of electromagnetic radiation. It is a chart that shows the entire "electromagnetic frequency spectrum." The electromagnetic spectrum includes cosmic rays that are radiated at a frequency of about 10^{22} Hz. (The number 10^{22} is a shorthand way to represent 1 followed by 22 zeros.) At the lower end of the radio portion of the chart, you see frequencies of 10^4 Hz (1 followed by 4 zeros, or 10,000 Hz).

Figure 13-1.
The Electromagnetic
Frequency Spectrum

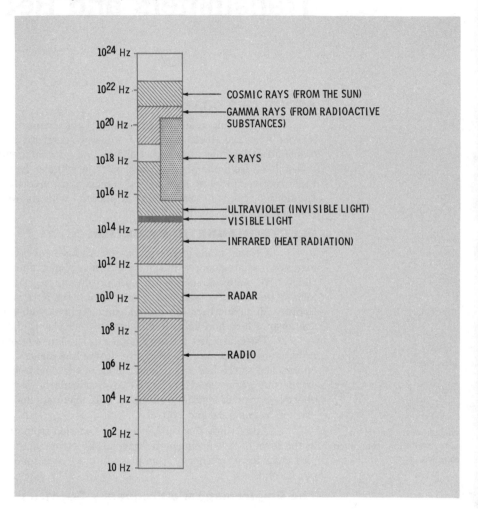

The Federal Communications Commission (FCC) has assigned specific groups of frequencies to different types of communications transmissions. This is called the "radio-frequency spectrum." *Figure 13-2* shows the details of that part of the electromagnetic spectrum. Commercial radio and television stations are permitted to radiate signals only in the portions of the radio-frequency spectrum that is assigned to them. Portions of the radio-frequency spectrum that are set aside for specific purposes are called bands.

The Assigned Broadcast Spectrum

The radio-frequency spectrum is a portion of the overall electromagnetic spectrum that is used for radio and television broadcasting.

**Figure 13-2.
The Radio-Frequency
Spectrum**

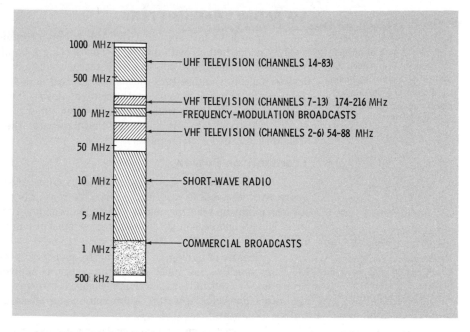

You are probably aware of the frequency assigned to your favorite local AM or FM radio station. *Figure 13-3* shows a typical dial for a portable AM/FM radio. Notice how the numbers on the dial relate to the bands of frequencies assigned to commercial stations by the FCC.

**Figure 13-3.
Typical AM/FM Radio
Receiver Dial**

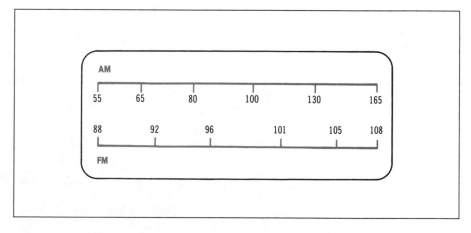

The dials on a commercial television receiver show channel numbers 1 through 83 rather than the actual frequencies assigned to them. However, the radio-frequency spectrum chart indicates the bands of frequencies set aside for those channels.

AM RADIO TRANSMITTERS

Each local radio station broadcasts at a different frequency. You select the station you want to hear by dialing the station's frequency on your radio receiver.

You have just seen that the AM dial on a commercial radio receiver is marked off with numbers representing a band of frequencies from 550 kHz through 1650 kHz. By rotating the tuning dial, you select the desired station. Because each local station broadcasts at a different frequency, you are able to select the one you desire. The same idea applies to an FM radio tuner as well as to a television tuner.

The device in a radio station that generates the energy that is broadcast is called a "transmitter."

Transmitter Power

You are aware that some stations come in stronger than others. There are two possible reasons for this. One reason is that the stronger stations might be transmitting at a higher power level than the weaker ones. The second reason is that the weaker station might be located farther away.

Broadcast power is measured in terms of watts or kilowatts (1 kilowatt equals 1000 watts).

Besides assigning an operating frequency to a radio station, the FCC also specifies how much power the station is allowed to use. Broadcast, or transmitter, power is measured in terms of watts—usually kilowatts (thousands of watts). Abbreviations for these units are W (watts) and kW (kilowatts).

The strength of a radio signal at your receiver is determined by the power of the broadcasting station and its distance from you.

Figure 13-4 shows two broadcast antennas transmitting at different frequencies in the broadcast band. Although one is farther away, its signal can reach the home because the station is using a higher-wattage transmitter. The signal from the lower-wattage station is able to reach the home only because it is a short distance away.

**Figure 13-4.
Transmission Power**

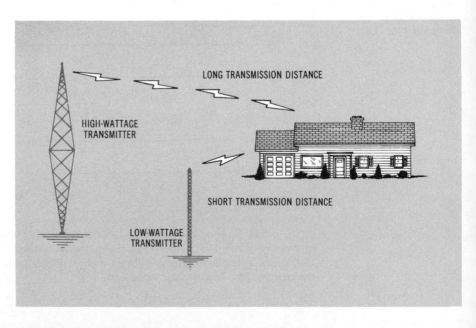

Carrier and Audio Frequencies

The frequency assigned to a radio station is called its "carrier frequency." The transmitter and its antenna are designed and tuned to operate at that specific frequency. As its name implies, the carrier frequency carries the reproduction of the sound originating in the studio. So there are actually two different frequencies involved in a radio transmission—the carrier frequency and the audio frequencies it carries.

A Basic AM Radio Transmitter

Figure 13-5 is a functional block diagram of a typical broadcast transmitter. It is called a functional block diagram because each block represents a different function in a very general way. The arrows indicate the direction of flow of the electronic processes.

**Figure 13-5.
Block Diagram of an
AM Radio Transmitter**

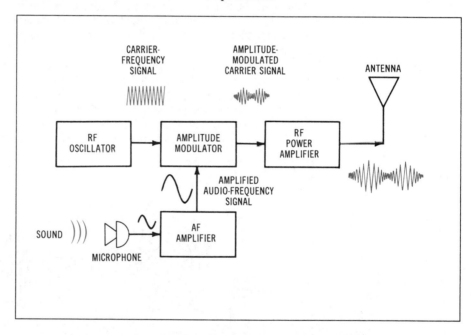

Sound enters the microphone and is fed to the "audio-frequency (af) amplifier." The signal coming directly from the microphone is an electrical version of the audio vibrations that enter it. These signals are too weak to be of any practical use, however; so it is necessary to amplify the microphone signal to a useful level. That is the main purpose of the af amplifier.

While the af amplifier is working the audio signal, the "radio-frequency (rf) oscillator" is generating a waveform that has the carrier frequency that is assigned to the station. The FCC specifies that this oscillator cannot vary more than 20 Hz above or below the assigned carrier frequency.

The audio signal from the af amplifier and the carrier signal from the rf oscillator meet at the "amplitude modulator." Here the audio signal is combined with the station's assigned carrier frequency to produce the complete broadcast signal. The audio frequency—always a much lower frequency—is combined with the carrier frequency in such a way that the audio frequency varies the amplitude of the carrier frequency.

The process of combining audio and radio-frequency signals is called modulation. An audio signal from the af section is said to modulate the carrier frequency signal. The modulation process illustrated here is called amplitude modulation (AM) because the audio-frequency signal varies the amplitude of the carrier frequency signal.

The amplitude-modulated waveform leaving the modulator has very little power—far too little to radiate energy more than a few feet. The purpose of the rf power amplifier is to boost the power to the hundreds of watts or kilowatts allowed for the station. This amplifier is simply a high-power amplifier that is designed to operate at the station's carrier frequency.

As illustrated in *Figure 13-6*, the AM signal from the transmitter is fed to an antenna. Power is fed to the antenna in the form of both current and voltage waveforms. The voltage portion sets up an electric field along the length of the antenna, and the current portion sets up a magnetic field. The result of this combination of electric and magnetic fields is a genuine electromagnetic wave that radiates outward at the speed of light.

> The process of combining audio- and carrier frequency signals is called modulation. Amplitude modulation is a modulation process whereby the audio signal varies the amplitude of the carrier signal.

> The antenna converts the voltage and current in the modulated waveform into an electromagnetic wave that radiates outward at the speed of light.

**Figure 13-6.
Antenna Radiation**

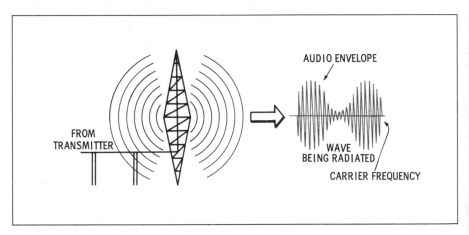

The frequency of the electromagnetic wave leaving the antenna is the same as the transmitter's carrier frequency. The audio information, modulated onto the carrier waveform, is carried along with the electromagnetic wave.

AM RADIO RECEIVERS

The purpose of a radio receiver is to intercept electromagnetic waves from a selected radio station, get rid of the carrier frequency portion of the signal, and reproduce the audio-frequency signal at a loudspeaker. In a manner of speaking, a radio receiver reverses the job performed by the transmitter.

Figure 13-7 is a simplified functional block diagram of an AM radio receiver. The antenna picks up the electromagnetic waves from all radio stations operating in the vicinity. The "tuner" lets you select the station you want to hear. The rf amplifier boosts the level of the modulated waveform, and the "detector" blocks the carrier portion and passes the audio portion to the af amplifier. The af amplifier boosts the amplitude of the audio signal to a level required for operating the loudspeaker.

**Figure 13-7.
Block Diagram of an
AM Radio Receiver**

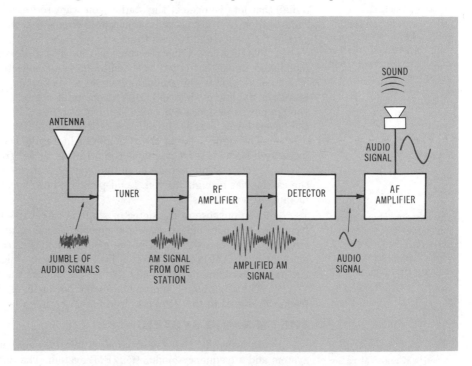

The electric and magnetic components of an electro-magnetic radio wave can induce voltage and current in a conductor.

Recall that radio waves are transmitted as electromagnetic waves—waves composed of electrical and magnetic fields. Those fields are capable of inducing a voltage and current in any conductor lying in their path. The frequency of the induced current and voltage is the same as the carrier frequency, and the amplitude of the induced current and voltage is proportional to the amplitude of the modulated signal.

A receiver antenna responds to the carrier frequencies of all stations having the strength to reach it.

A receiver antenna is a conductor that is intentionally placed within the electromagnetic radio waves and specifically designed to respond to them. The voltage and current thus induced in the antenna are extremely low-power versions of the signals applied to the antenna at the transmitter. Unlike a transmitter antenna that operates at a single frequency (the station's carrier frequency), a receiver antenna is filled with a multitude of frequencies from different radio stations.

If you were to convert all of the signals on a receiver antenna into audio sounds, you would hear a meaningless jumble of a large number of radio stations. It is thus necessary to sort out the one you want to hear and block the remainder of them. That is the main job of the tuner section of your radio receiver.

The tuner section lets you adjust a circuit that passes the carrier frequency of the station you want to hear and rejects all other frequencies from the antenna.

The tuner section of a radio receiver has a manual control (usually a dial) that lets you select the station you want to hear. When you tune in a radio station, you are actually adjusting a circuit that passes that station's carrier frequency and rejects the rest.

The signal leaving the tuner is thus a local version of the modulated signal that is fed to the antenna at the selected radio station. This local version of the modulated signal is very weak, however. It is often less than 100 microvolts (μV) (1 μV is one-millionth of a volt). That is hardly enough power to operate a typical loudspeaker, so it is necessary to boost the power level. That is the purpose of the rf amplifier in a radio receiver.

The signal leaving the rf amplifier in a radio receiver is a modulated signal. Its frequency is equal to the carrier frequency of the selected radio station, and its amplitude varies with the audio signal that the carrier has brought to you. Loudspeakers, however, do not respond to modulated waveforms.

The detector stage blocks the carrier frequency and passes the audio-frequency signal to the remaining circuitry.

The purpose of the detector stage is to get rid of the carrier frequency and pass only the audio-frequency signal to the remaining circuitry. It reverses the task of the modulator in the transmitter. The detector in an AM receiver is often little more than a high-frequency rectifier (a germanium diode) and a filter that blocks the carrier frequency.

The af amplifier simply boosts the power level of the audio-frequency signal to the level required for operating a loudspeaker.

THE FM RADIO SYSTEM

The main difference between AM and FM radio is the way in which the audio-frequency signal is modulated with the carrier-frequency signal.

In principle, there is only one difference between an AM radio system and a frequency-modulation (FM) system. The difference is in the way the audio-frequency waveform is modulated with the carrier frequency waveform. There are vast differences between the ways the FCC allows commercial AM and FM stations to handle their signals, but these differences have little to do with the technology itself.

An AM signal has a constant frequency and a varying amplitude. An FM signal has a constant amplitude and a varying frequency.

Figure 13-8 illustrates the primary difference between AM and FM signals. You have already seen that the audio-frequency waveform modulates the carrier frequency of an AM signal by varying the amplitude of the carrier. That is why it is called amplitude modulation. By contrast, the audio-frequency signal in an FM system modulates the carrier by shifting its frequency. Notice that an AM signal has a constant frequency but a varying amplitude. An FM signal has a constant amplitude but a varying frequency.

**Figure 13-8.
The Primary Difference
Between AM and FM**

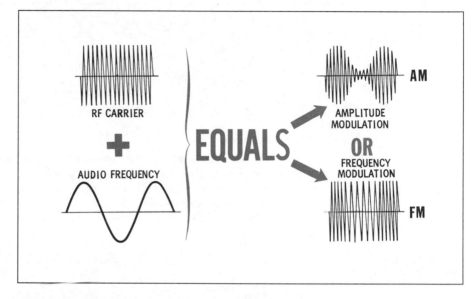

FM Transmitters

The rf oscillator in an FM transmitter generates the carrier frequency, or center frequency, assigned to the station.

Figure 13-9 is a simplified block diagram of an FM radio transmitter. The af amplifier boosts the level of the signal from the microphone, and the rf oscillator generates the carrier frequency waveform assigned to the station. Because the modulator varies that frequency below and above the assigned carrier frequency in an FM system, the assigned frequency is often called the "center frequency."

The modulator in an FM transmitter varies the carrier frequency according to the amplitude of the audio-frequency signal.

The purpose of the frequency modulator is to modulate the audio-frequency signal with the station's center frequency. The result is a carrier frequency that varies above and below the center frequency, depending on the amplitude of the audio-frequency waveform. The greater the amplitude of the audio-frequency waveform, the greater the frequency deviation from the center frequency. An FM signal thus carries audio information as changes in the carrier frequency.

The rf power amplifier boosts the FM waveform to the power level allowed by the FCC. The antenna converts the voltage and current components of the FM waveform to electromagnetic waves that radiate from the antenna at the speed of light.

Figure 13-9.
Block Diagram of an
FM Radio Transmitter

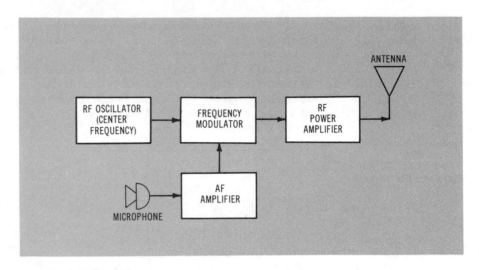

FM Receivers

Figure 13-10 is a block diagram of a basic FM radio receiver. The antenna converts all radio radiation impinging upon it to voltage and current waveforms. The tuner allows you to tune in the center frequency of the station you want to hear. The signal leaving the tuner is a low-power version of the signal that is fed to the antenna of the station you have selected.

Figure 13-10.
Block Diagram of an
FM Radio Receiver

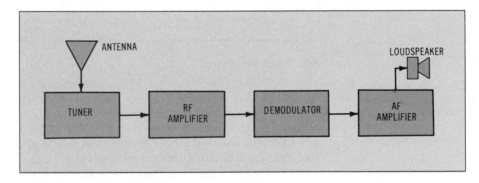

In an FM receiver, the demodulator converts the changes in carrier frequency to the corresponding audio-frequency waveform.

The rf amplifier boosts the level of the selected FM signal before applying it to the "demodulator." The demodulator in an FM receiver is the portion of the circuit that undoes the work of the frequency modulator in the transmitter—it converts the changes in frequency to an audio-frequency waveform. The af amplifier boosts the audio signal to the level required for operating the loudspeaker.

DIFFERENCES BETWEEN AM AND FM RADIO SYSTEMS

You have already learned that the primary technical difference between AM and FM radio systems is the method used for modulating the audio-frequency signal onto the radio-frequency carrier signal. There are some other differences that are not quite as fundamental in nature but nevertheless important to your understanding of commercial radio.

FM Broadcasts Have Less Noise

Consider that an AM waveform has a fixed frequency and carries its audio information in the form of changes in amplitude. Ideally, there is nothing between the transmitter and receiver that causes unwanted changes in the amplitude of the signal. But the real world is far from ideal. *Figure 13-11* shows how electrical static affects the amplitude of an AM waveform. Notice how electrical interference, such as that from lightning or nearby electrical machinery, can add spikes onto an AM signal during transmission. The radio receiver cannot distinguish the spikes from the desired audio signal, so these spikes emerge from the loudspeaker as static noise.

**Figure 13-11.
Unwanted Noise on AM
and FM Transmissions**

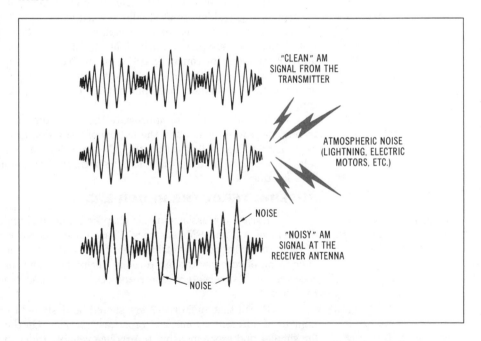

The audio signal from an FM receiver is relatively free of static.

FM signals are not immune to such interference. Bear in mind, however, that the desired audio information in an FM system is carried as changes in frequency. The detector section in an FM receiver ignores changes in amplitude and responds only to changes in frequency. Any interference riding on an FM transmission is thus ignored and is barely reproduced (if at all) at the loudspeaker. This accounts for the static-free reception of an FM receiver compared with that of an AM receiver.

FM Broadcasts Have Higher Fidelity

Another difference between commercial AM and FM radio systems is that FM reproduces a much greater range of audio frequencies (high fidelity) and has been used for stereophonic broadcasts for a number of years. AM signals, on the other hand, do not reproduce the higher audio frequencies, and the technology for AM stereophonic broadcasting is still in its infancy.

The limitations of AM broadcasting are imposed by early FCC legislation rather than the state of modern radio technology.

There is nothing inherent in the FM system that gives it higher fidelity than AM broadcasts. And there is nothing about the nature of AM that prevents it from working with a much wider range of audio frequencies. These differences are imposed by the history of FCC legislation. AM radio was invented first, and the FCC allocated the frequency bands for commercial AM broadcasting on the basis of a primitive radio technology. The die has been cast, and it would be too costly to change things. So in spite of tremendous advances in electronics and AM technology, broadcast stations are bound to the limitations of an outmoded way of doing things.

Commercial FM broadcasting offers higher fidelity and other features because the FCC legislation was passed at a time when radio technology was advancing at a rapid pace.

On the other hand, FM technology came along at a time when electronics was making great strides toward working at higher frequencies. Notice from the radio-frequency spectrum in *Figure 13-2* that commercial FM stations are assigned a much higher band of frequencies. The FCC set up its regulations for commercial FM on the basis of a superior technology. So it is little wonder that FM broadcasts offer higher quality and more opportunities for further development than commercial AM broadcasting can.

You will be able to appreciate the full potential of AM transmission when you study the nature of television broadcasting in the next chapter. You will find that the AM technology used for transmitting television images is far superior to that permitted for commercial FM radio broadcasting.

SUPERHETERODYNE RECEIVERS

Most radio (and television) receivers are superheterodyne receivers.

Earlier discussions in this chapter included simplified block diagrams of AM and FM receivers. These diagrams portray the basic ideas of radio receivers but do not include a feature that is common to just about every kind of receiver produced today (including television receivers). Most receivers are "superheterodyne receivers," also called "superhet" receivers.

Receiver circuits designed to operate at just one frequency are simpler, less costly, and more reliable than circuits designed to operate at many different frequencies.

Recall that radio receivers should be designed to receive a large number of different commercial broadcast stations. On the other hand, it is far simpler and less expensive to produce reliable radio circuits that tune just one frequency.

The first AM radio receivers, for example, were difficult to tune. Rather than being able to turn a single dial for selecting the desired station (carrier frequency), you had to tune each rf stage individually—and there were often three or more stages that had to be tuned in order to get the best possible reception.

A superheterodyne frequency converts the selected broadcast frequency to the same frequency for processing through the remainder of the rf sections. This one frequency is called the intermediate frequency.

The superheterodyne principle solves these difficulties by converting any selected carrier frequency to the same frequency for processing through the rf sections of the receiver. The antenna and tuner are the only sections that have to be designed to operate at any broadcast frequency. The remaining rf stages operate at the same frequency, called the "intermediate frequency (if)."

Figure 13-12 is a block diagram of a superheterodyne receiver. The same general diagram applies to AM and FM receivers. The diagram differs from those of the simpler AM and FM receivers by the addition of two stages, the "local oscillator" and "mixer," and the replacement of an rf amplifier with an if amplifier. The dashed line running between the tuner and local oscillator indicates that they are mechanically ganged together so that when you adjust the tuner you also adjust the local oscillator.

**Figure 13-12.
Block Diagram of a
Superheterodyne
Receiver**

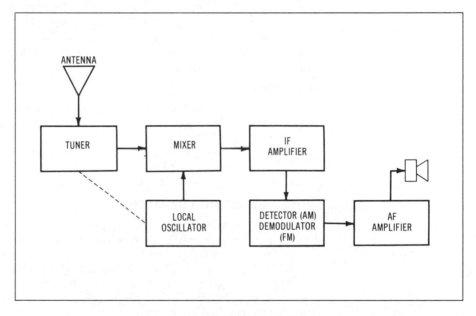

The antenna responds to all electromagnetic radio waves, generating microwatts of power for the tuner. The tuner circuit can be manually adjusted to pass only the carrier frequency of the selected broadcast station.

The tuner and local oscillator are ganged together so that the difference between the selected carrier frequency and the local oscillator frequency is always the same.

The local oscillator produces a frequency that is proportional to the selected carrier frequency. As you adjust the tuner to select different carrier frequencies, you also adjust the frequency of the local oscillator as well. The tuner and local oscillator are designed so that the difference between their two frequencies is always the same. In most commercial AM radio receivers, the difference between the selected broadcast carrier frequency and the local oscillator frequency is 455 kHz (455,000 Hz).

The selected broadcast carrier frequency and the local oscillator frequency are combined in the mixer stage. The result is the receiver's intermediate frequency. The intermediate frequency is always the same frequency, regardless of the frequency of the selected broadcast station.

The signal from the tuner and the signal from the local oscillator are combined in the mixer stage (sometimes called the first detector). The output of the mixer stage is the difference between the two frequencies. This "difference frequency" is the receiver's intermediate frequency. In a properly designed superheterodyne receiver, the intermediate frequency is always the same, regardless of what broadcast carrier frequency you choose. Recall that commercial AM radio stations are assigned a carrier frequency anywhere between 550 kHz and 1650 kHz. No matter which you choose, the if signal from the mixer is at 455 kHz.

The if amplifier simply boosts the power level of the received AM or FM signal. Bear in mind, however, that it has to work only with the single intermediate frequency. Modern receivers often contain at least two if amplifiers arranged one after the other.

The if amplifier sends its signal to a detector or demodulator, depending on whether it is an AM or FM receiver. In either case, this section recovers the audio-frequency portion of the signal and gets rid of the if portion. The af amplifier simply boosts the power level of the audio signal for operating the loudspeaker.

TRANSCEIVERS

A "transceiver" is a radio device that combines a transmitter and a receiver. Amateur ("ham") radio CB (citizen's band) rigs are popular examples of transceivers. Transceivers, or two-way radios, play vital roles in aircraft and marine (shipboard) communications. No matter what they are called or how they are used, the basic principle is the same: they combine the operations of a radio transmitter and receiver.

WHAT HAVE WE LEARNED?

1. The electromagnetic spectrum represents all forms of radiant energy, including radio wave, infrared and ultraviolet light, visible light, X rays, gamma rays, and cosmic rays.
2. Radio (and television) transmissions occupy a portion of the electromagnetic spectrum called the radio-frequency (rf) spectrum.
3. Electromagnetic waves, including radio waves, radiate outward at a speed of 186,000 miles per hour.
4. The Federal Communications Commission is responsible for assigning the operating frequencies and power limitations to U.S. commercial radio and television stations.
5. The commercial AM broadcast band is between 550 kHz and 1650 kHz. The commercial FM broadcast band is between 88 MHz and 108 MHz.
6. Broadcast power is expressed in terms of watts and kilowatts.
7. The strength of a broadcast signal at a receiver is mainly determined by the power rating of the station and its distance from the receiver.
8. The operating frequency of a transmitter is its carrier frequency.
9. An audio-frequency (af) amplifier boosts the power level of an audio electrical signal. In a transmitter, af amplifiers amplify the audio signal from the microphone. In a receiver, af amplifiers boost the audio-signal power to operate the loudspeaker.

10. The rf oscillator in an AM transmitter generates the carrier frequency. The rf oscillator in an FM transmitter generates the center frequency.
11. The process of "attaching" audio information to a carrier frequency is called modulation.
12. In an amplitude-modulation (AM) system, audio information is carried by changes in the amplitude of the carrier signal.
13. In a frequency-modulation (FM) system, audio information is carried by changes in the frequency of the carrier signal.
14. An AM signal has a constant frequency but a changing amplitude. An FM signal has a constant amplitude but a changing frequency.
15. The antenna at a transmitter changes electrical current and voltage into electromagnetic waves. The antenna at a receiver changes electromagnetic waves into electrical current and voltage.
16. The tuner section of a radio receiver selects the carrier frequency of the desired broadcast station.
17. The detector stage in an AM receiver separates the audio-frequency signal from the radio-frequency carrier. This stage in an FM receiver is called the demodulator.
18. FM reception is relatively immune to electrical static that can attach itself to the signal during transmission.
19. Most of the differences between AM and FM broadcasting are due more to legislation than technical capabilities.
20. Commercial FM stations can broadcast high-fidelity stereo information because the FCC has allocated sufficiently wide bands of frequencies for the purpose.
21. Most radio and television receivers are superheterodyne, or "superhet," receivers.
22. The main advantage of a superheterodyne receiver is that most stages can be designed to operate at a single frequency called the intermediate frequency.
23. The mixer, or first detector, in a superheterodyne receiver combines the carrier frequency from the selected station with the local oscillator frequency. The result is the receiver's intermediate frequency.
24. A transceiver is a radio device that combines transmitter and receiver circuits into a single piece of equipment.

KEY WORDS

Amplitude modulation (AM)
Amplitude modulator
Antenna
Audio-frequency (af) amplifier
Carrier
Carrier frequency
Center frequency
Demodulator
Detector
Electromagnetic frequency spectrum
Electromagnetic waves
Federal Communications
  Commission (FCC)
First detector

Frequency modulation (FM)
Frequency modulator
Intermediate frequency (if)
Intermediate frequency (if) amplifier
Local oscillator
Mixer
Modulator
Radio-frequency (rf) oscillator
Radio-frequency (rf) spectrum
Radio-frequency (rf) power amplifier
Superheterodyne receiver
Transceiver
Transmitter
Tuner

Quiz for Chapter 13

1. Which one of the following statements is true?
   a. Radio waves travel at a much slower speed than light waves.
   b. Radio waves can travel only through the atmosphere, whereas light waves can travel through space as well as the atmosphere.
   c. Radio waves have higher frequencies than light waves.
   d. Radio waves and light waves are simply different parts of the same electromagnetic spectrum.

2. If you find a commercial AM station at 101 on your AM dial, its assigned carrier frequency must be:
   a. 101 Hz.
   b. 101 W.
   c. 101 kHz.
   d. 101 kW.
   e. 10.1 MHz.
   f. 101 MHz.

3. A commercial FM station that advertises itself as "97 FM" most likely has an assigned center frequency of:
   a. 97 Hz.
   b. 97 W.
   c. 97 kHz.
   d. 97 kW.
   e. 97 MHz.
   f. 970 MHz.

4. Which part of a transmitter generates the carrier frequency?
   a. Rf oscillator.
   b. Rf amplifier.
   c. Af amplifier.
   d. Modulator.
   e. Antenna.
   f. Detector.

5. Which part of a transmitter amplifies the signal from the microphone?
   a. Rf oscillator.
   b. Rf amplifier.
   c. Af amplifier.
   d. Modulator.
   e. Antenna.
   f. Detector.

6. Which part of a transmitter system converts the current and voltage components of the modulated signal into electromagnetic waves?
   a. Rf oscillator.
   b. Rf amplifier.
   c. Af amplifier.
   d. Modulator.
   e. Antenna.
   f. Detector.

7. Which part of an AM radio transmitter mixes the audio and carrier-frequency signals?
   a. Rf oscillator.
   b. Rf amplifier.
   c. Demodulator.
   d. Modulator.
   e. Antenna.
   f. Detector.

8. Which section of an AM radio receiver is mainly responsible for separating the carrier signal of the desired station from all others?
   a. Antenna.
   b. Tuner.
   c. Detector.
   d. Mixer.
   e. Af amplifier.
   f. Rf amplifier.

9. Which section in an AM radio receiver is mainly responsible for separating the audio signal from the modulated rf signal?
   a. Antenna.
   b. Tuner.
   c. Detector.
   d. Mixer.
   e. Af amplifier.
   f. Rf amplifier.

10. Which one of the following statements most accurately describes the main technical difference between the AM and FM signals?
    a. FM stations broadcast in a higher frequency band than AM stations do.
    b. FM stations are capable of broadcasting high-fidelity, stereophonic sounds; AM stations cannot.
    c. FM stations are not required to broadcast commercials; AM stations must carry commercials.

    d. AM signals have a fixed amplitude but varying frequency; FM signals have a varying amplitude but a fixed frequency.
    e. FM signals have a fixed amplitude but varying frequency; AM signals have a varying amplitude but a fixed frequency.

11. What is the unique feature of superheterodyne receivers?
    a. They are intended only for receiving FM broadcasts.
    b. They convert all carrier frequencies to a single if signal.
    c. They mix the incoming carrier frequency with that of a local oscillator to separate stereophonic channels.
    d. They do not require an antenna.

Understanding the Television System

ABOUT THIS CHAPTER

In this chapter you will learn how a television transmitter generates picture signals, and you will see that the sound portion of the signal is simply an FM broadcast at a slightly different frequency. You will gain more knowledge of antennas and the problems of transmitting electromagnetic waves. You will also become familiar with how a television receiver converts electronic signals into pictures. Before starting your work in this chapter, make certain you understand how a CRT works (Chapter 11) and how audio signals are transmitted and received (Chapter 13).

THE VIDEO SIGNAL

Figure 14-1 shows the main elements of a standard television video signal. As you learned in Chapter 11, this signal consists of video information, blanking pulses, and synchronizing pulses. The video information represents the image to be drawn on a CRT, the blanking signals turn off the electron beam during the horizontal and vertical retrace intervals, and the synchronizing pulses trigger the start of horizontal and vertical sweep operations.

**Figure 14-1.
Elements of a Video
Signal**

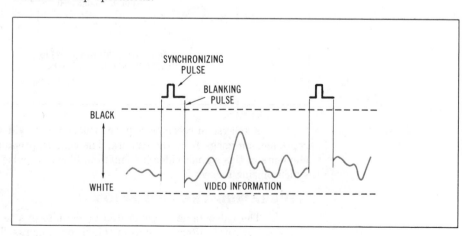

Magnetic deflection coils around the neck of a CRT move the beam in response to signals applied to them. A sawtooth waveform applied to the horizontal deflection coil moves the beam from left to right on the slowly rising portion of the sawtooth. Likewise, a sawtooth waveform applied to the vertical deflection coil moves the beam from top to bottom on the slowing rising portion of the sawtooth. The purpose of the synchronizing pulses in the video signal is to make sure the horizontal and vertical sweep waveforms begin at the proper time.

Figure 14-2 shows that a sawtooth waveform is characterized by a relatively slow rise in voltage followed by a sudden drop to the baseline. As just described, the beam is swept from left to right across the screen during the slowly rising portion of the horizontal sawtooth and from top to bottom during the rising portion of the vertical sawtooth. During the time a sawtooth waveform is dropping to its baseline voltage, the beam is snapped back to the point on the screen where the next scan is to begin. The beam must be turned off during this retrace interval. This is the purpose of the blanking pulses.

**Figure 14-2.
Elements of Sweep
Waveforms**

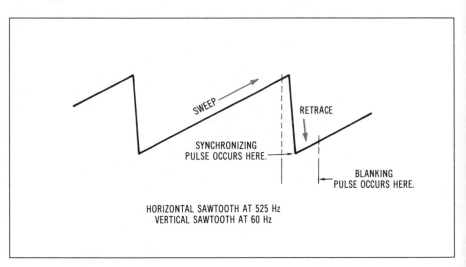

A television receiver requires the complete video signal in order to reproduce an image from the camera. The camera produces all three elements of this signal: video information, blanking pulses, and synchronizing pulses.

GENERATING THE VIDEO SIGNAL

The video signal is generated by circuits in a television camera. There are several different types of television cameras. The iconoscope, image dissector, and "image orthocon" are examples. The last is the type most often used for commercial television broadcasting. Although the manner in which the various types accomplish their purpose differs, their basic operating principles are the same.

The target in a camera tube is made of a material that transforms areas of light and dark into corresponding areas of electrical charges.

As shown in *Figure 14-3*, a television camera focuses the light from a scene through a lens and onto a target within the camera. This is the same principle used by ordinary photographic cameras; however, the target in a television camera is made of a light-sensitive material that generates areas of electrical charges proportional to the intensity of the light. The optical image on the target of this television camera is thus reproduced as areas of electrical charges.

**Figure 14-3.
Camera Operating
Principle**

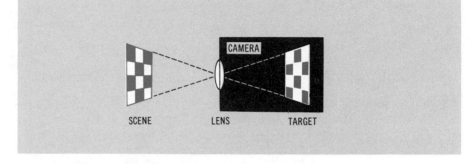

The optical image and its pattern of electrical charges are on the face of a camera tube. A camera tube is a vacuum tube, and it works much like a CRT; but instead of using a beam of electrons to generate patterns of light and dark on the screen, a camera tube uses a beam of electrons to determine the patterns of light and dark for the image focused on it.

A television camera uses the same beam-deflection patterns as used by a television receiver.

Figure 14-4 shows that a television camera uses the same beam-deflecting patterns as used by a television receiver. Horizontal deflection waveforms sweep the camera beam from left to right across the image, and vertical deflection waveforms move the beam downward. The scanning pattern for a television camera is exactly the same as the raster-scan pattern used for television receivers.

A television camera works much like a CRT in reverse.

A television camera generates the video information by scanning an electron beam across the image focused on the screen. This image creates areas of electrical charges that, in turn, change the amount of current in the electron beam. So a television camera works much like a CRT in reverse. Changing the amount of electron-beam current in a CRT produces changes in the amount of light produced on the screen. Changing the brightness of the light falling on the target of a camera tube produces changes in the amount of electron-beam current.

**Figure 14-4.
Sweep Patterns for TV
Cameras**

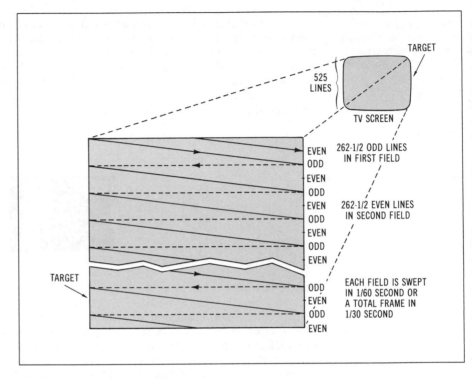

Figure 14-5 is a functional block diagram of a television camera system. A master oscillator produces very precise timing pulses that are used by the sweep generators and sync-and-blanking pulse generators. The horizontal and vertical sweep generators produce the sawtooth waveforms that are required for sweeping the electron beam in the camera tube through its left-to-right, top-to-bottom pattern. The video mixer adds the synchronizing and blanking pulses to the beam-current signal.

Color television cameras use a separate camera tube for red, green, and blue. These three beam-current signals are combined into two kinds of video information: luminance information and chrominance information. The "luminance" portion of the video information is identical to the video information for a black-and-white transmission—it carries brightness information. The "chrominance" part of the signal carries the color information. Chrominance information is added to the video signal as a "color burst signal" that appears during the horizontal blanking interval, just after the horizontal synchronizing pulse.

Color television cameras combine beam currents from red, green, and blue tubes into luminance and chrominance signals.

TRANSMISSION OF VIDEO AND AUDIO SIGNALS

The video signal from a television camera is a complex waveform that combines video information, blanking pulses, and synchronizing pulses. All of this information occupies a band of frequencies from 0 Hz to 4.5 MHz. The video signal for a single television station occupies more space in the radio-frequency spectrum than the entire commercial AM radio band.

**Figure 14-5.
Block Diagram of the
TV Camera System**

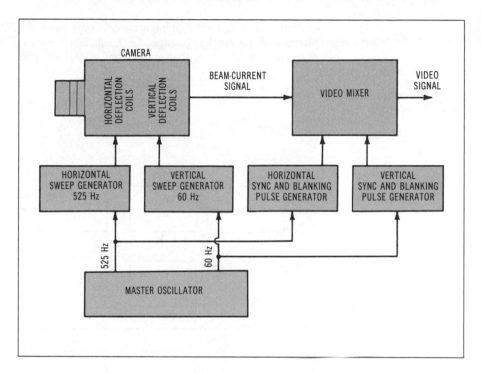

Video information is
transmitted as an AM signal. The accompanying audio information is
transmitted as an FM signal. The sound signal is
4.5 MHz above the carrier
frequency for the station.

At the television transmitter, this wideband video signal is
amplitude-modulated with the station's assigned carrier frequency. The
audio signal accompanies the video information, but the audio is frequency-
modulated on a carrier that is 4.5 MHz higher than the video carrier
frequency.

A modulated television signal that carries all video information,
sync and blanking pulses, and sound is called a "composite video signal."
The composite video signal is amplified to a power level allowed by the
station's license. It is then sent to the antenna, where it is converted into
electromagnetic waves that can be broadcast.

TELEVISION RECEIVERS

A television receiver is basically a superheterodyne AM/FM
receiver. When you select a channel for viewing, you actually tune the
carrier frequency for the video portion of the signal. Subsequent circuits
separate the AM video information from the FM sound signal. The video
signal is further separated into signals for luminance (including the blanking
pulses), chrominance (if you are using a color television), and the
synchronizing pulses.

Figure 14-6 shows the first main sections of a television receiver.
The tuner includes the familiar channel selector for commercial VHF (very-
high-frequency) and UHF (ultra-high-frequency) television stations. Each
channel number represents a broadcast frequency that is assigned to the

station by the Federal Communications Commission (FCC). The main purpose of the tuner is to separate the carrier frequency of the selected channel from all others that are present at the antenna.

**Figure 14-6.
Rf, If, and Af Sections
of a TV Receiver**

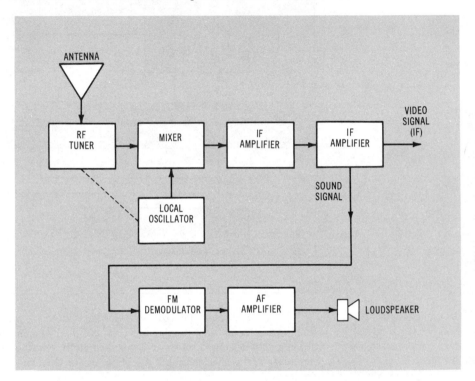

Television receivers all use the superheterodyne principle described in the previous chapter. This involves the rf tuner, local oscillator, and mixer stages of the receiver. As you use the tuner to select a channel, you are also changing the frequency of the local oscillator. The two frequencies—the selected channel and the local oscillator frequencies—are combined in the mixer to produce the receiver's intermediate frequency (if). No matter what channel you choose to view, it always leaves the mixer stage with the same frequency.

Television receivers use at least two if amplifier stages in order to boost the signal level to a useful amplitude. The final if amplifier includes a tuning circuit that separates the sound signal from the composite video signal. Bearing in mind that the sound portion of the signal is frequency modulated, you can see that there is a need for an FM demodulator to convert the changes in frequency to changes in amplitude. The audio-frequency (af) signal is then further amplified before it is applied to the loudspeaker.

Figure 14-7 is a block diagram of the video section in a typical television receiver. At this point in the circuit, the composite video is being carried by the if. This if must be discarded before the video signal can be properly handled by the remaining portions of the system. The signal is amplitude-modulated, so it is appropriate to use an AM detector circuit to remove the if carrier frequency. This is the job of the video detector.

**Figure 14-7.
The Video, Sync, and
Sweep Sections of a TV
Receiver**

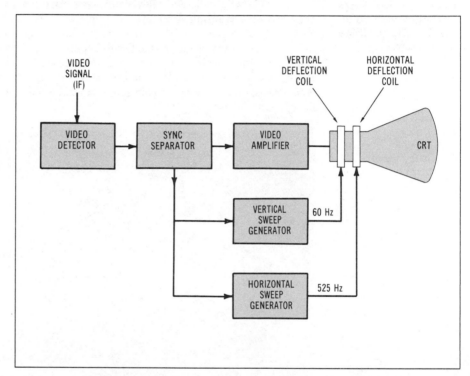

The demodulated video signal goes to a circuit that separates out the synchronizing pulses. This section is appropriately called the sync separator. The sync pulses are sent to the vertical and horizontal sweep generators, and the remainder of the video signal goes to a video amplifier.

The vertical and horizontal sweep generators produce the sawtooth waveforms that are required for operating their respective magnetic deflection coils. These generators operate at about 60 Hz (vertical) and 525 Hz (horizontal) as long as the receiver is turned on. This way you can see a raster on the screen, even when you are not tuned to a station. But when you tune in a station, the sync pulses lock the sweep generators into step with the corresponding sweep generators way back in the station's camera.

The video amplifier section applies the video information and blanking pulses to the CRT. These signals vary the amount of beam current and, as a result, the patterns of light and dark on the face of the CRT.

The video section in a color television receiver is a bit more complicated. The chrominance signal has to be removed from the tops of the blanking signals, demodulated, and transformed into three separate video signals for red, green, and blue.

TRANSMITTING TELEVISION SIGNALS

Lower-frequency electro-magnetic waves tend to hug the earth and follow its curvature.

The way radio and television waves travel, or propagate, once they leave the transmitter antenna changes with their frequency. *Figure 14-8* shows that waves of lower frequency tend to hug the earth and travel along its curvature. Some of this lower-frequency energy moves away from the earth, but some of that energy is reflected back to the earth by the ionosphere. Commercial AM radio broadcast stations operate at these frequencies, and this may explain some unusual effects you might have observed at times.

**Figure 14-8.
Propagation of Lower-Frequency Waves**

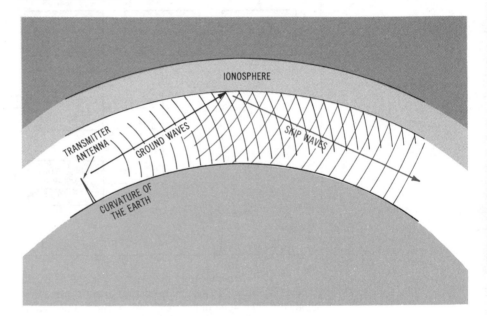

Consider a relatively powerful AM radio station operating in San Francisco. Because its waves tend to hug the earth, you will be able to receive the station reliably well over the horizon—maybe up to a radius of several hundred miles. Under a very special set of conditions, however, you might be able to receive that same station as far away as Chicago, but not in Denver (which is much closer to San Francisco). How is this possible? The answer is that the ionosphere is reflecting signals from the San Francisco station back to earth in the area of Chicago, and it skipped completely over Denver. This "skip effect" is much the same thing as a mirage you might see in the desert. This effect is not reliable, but it does prove that electromagnetic waves can be reflected back to earth.

Commercial FM radio and VHF/UHF stations broadcast at relatively high frequencies. As illustrated in *Figure 14-9*, higher-frequency waves tend to travel in straight lines. They do not follow the curvature of the earth, and a great deal of their energy bores right through the ionosphere and into space. As a result, these stations cannot be received very far beyond the horizon. This is often called "line-of-sight propagation." The skip phenomenon does occur for the lower-frequency VHF television channels, but it is extremely rare.

**Figure 14-9.
Propagation of Higher-
Frequency Waves**

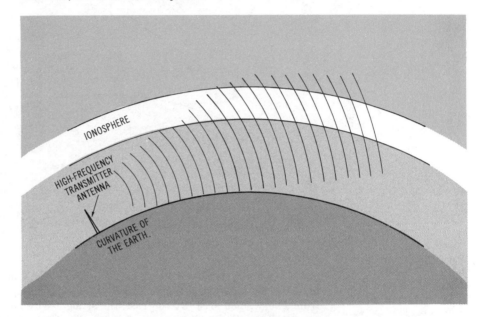

One way to cope with the line-of-sight problem with VHF/UHF television broadcasting is to locate the antenna on top of a very tall building. This procedure extends the range of the horizon and, therefore, the broadcast range.

Communications Satellites

A communciations satellite located in a fixed position over the earth is like a line-of-sight antenna that is hundreds of miles tall.

Modern satellite technology gives all broadcast stations the potential for unlimited range. A space satellite has an earth horizon of thousands of miles. It is like a line-of-sight antenna that is hundreds of miles tall. The signal from a broadcast station is intentionally directed to a space satellite that is in a fixed location over the earth. The signal is then rebroadcast to the earth from this "communications satellite." Any ground station that can "see" the satellite can receive its signals.

Broadcast signals are sent to the satellite and returned to earth in a band of frequencies that is different—much higher—than the carrier frequencies for commercial broadcasts. This is necessary so that a single satellite can handle hundreds of stations at the same time without any two stations using the same frequency. So you cannot receive satellite rebroadcasts with the usual television receiver system. *Figure 14-10* shows the basic elements of a satellite receiver system.

**Figure 14-10.
Elements of a Satellite
Receiver System**

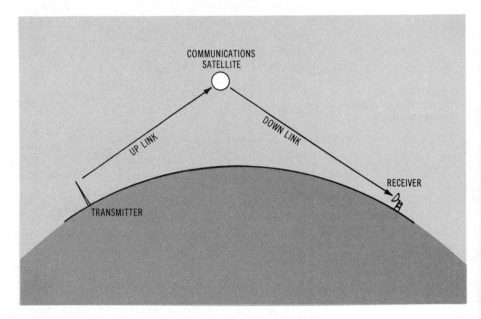

Parabolic, or "dish," antennas are highly directional and must be aimed directly at the communications satellite you want to use.

Parabolic Antennas

First, you must have a "parabolic," or "dish," antenna. Parabolic antennas are highly directional. That is, they can pick up signals coming from a very small area. By contrast, the straight-wire antenna used for automobile radios receives signals from all directions at the same time. Parabolic antennas must be aimed precisely at the communications satellite you want to use. This eliminates a lot of other signals it might receive from other earth-bound stations and satellites.

Your satellite communications system must also have a special low-noise "preamplifier." The preamplifier is usually located at the focal point of the parabolic antenna. Its purpose is to boost the signal level before it can be messed up with static noise and other kinds of electromagnetic interference in the area of your system.

You also need a special tuner. As mentioned earlier, satellite communications systems use hundreds of different channels at frequencies far above those allocated for normal television broadcasting. This tuner converts the incoming signals to frequencies that are compatible with the tuner in your ordinary television receiver.

Cable Television Systems

Satellite communications systems were not originally intended for direct use by the general public. Commercial satellite communications are meant to be used by cable television systems. The general idea is to use the satellite system to provide a choice of hundreds of commercial television channels for cable television companies. The cable companies select the stations and send them to individual subscribers through a network of cables.

Because cable signals are confined to a wire, the signal applied to the television receiver is not prone to atmospheric interference. Confining the broadcast to a cable eliminates the need for expensive rf power amplifiers and antennas at the station. And because the signal is not broadcast into the public domain, cable stations are not subject to many FCC rules that govern the quantity and quality of programming that is broadcast by usual means.

WHAT HAVE WE LEARNED?

1. Television camera tubes and the CRT in a television receiver use the same raster-scan patterns.
2. Synchronizing, or "sync," pulses determine when the horizontal and vertical beam-deflecting waveforms begin.
3. Blanking pulses turn off the electron beam during the horizontal and vertical retrace intervals.
4. The most common kind of television camera tube is the image orthocon.
5. The target in a television camera tube is made of a material that changes patterns of light and dark light into corresponding patterns of electrical charges.
6. The pattern of electrical charges on the target of a television camera tube changes the amount of current in the electron beam.
7. Color television cameras use three different tubes: one for red, one for green, and one for blue.
8. Color television cameras generate two video elements: the luminance (brightness) information and the chrominance (color) information.
9. Television video is transmitted as an AM signal; the sound is transmitted as an FM signal.
10. The sound portion of the broadcast is transmitted at 4.5 MHz above the station's assigned carrier frequency.
11. A fully modulated television signal, including video information, sync and blanking pulses, and sound, is called a composite video signal.
12. The channel selector on a television receiver is actually the tuner that separates the carrier frequency of the desired station from all others.
13. Television receivers are superheterodyne receivers, so you can expect to find a local oscillator, mixer, and if amplifiers.
14. The sound portion of the composite video signal is separated from the video portion, demodulated as an FM signal, and amplified before it is applied to a loudspeaker.
15. A sync separator circuit sorts the sync pulses from the video signal and applies them to the vertical and horizontal sweep generators built into the television receiver.
16. Lower-frequency electromagnetic waves tend to hug the earth and propagate some distance over the horizon.

17. Higher-frequency electromagnetic waves travel in straight lines and propagate into space.
18. A communciations satellite located in a fixed position over the earth is like a line-of-sight antenna that is hundreds of miles tall.
19. Parabolic, or "dish," antennas are highly directional.

KEY WORDS

Chrominance signal	Preamplifier
Color burst signal	Propagate
Communications satellite	Skip effect
Image orthocon	Target
Line-of-sight transmission	Ultra-high frequency (UHF)
Luminance signal	Very-high frequency (VHF)
Parabolic antenna	Video detector

Quiz for Chapter 14

1. Which one of the following waveforms in a television system is responsible for cutting off the electron beam during the horizontal and vertical retrace intervals?
   a. Synchronizing waveform.
   b. Blanking pulses.
   c. Sweep waveform.
   d. Sound waveform.

2. Which one of the following waveforms in a television system is most directly responsible for moving the electron beam in a camera tube or CRT?
   a. Synchronizing waveform.
   b. Blanking pulses.
   c. Sweep waveforms.
   d. Sound waveforms.

3. Which one of the following waveforms is applied to the deflection coils of a television camera tube or CRT?
   a. Synchronizing waveform.
   b. Blanking pulses.
   c. Sweep waveforms.
   d. Sound waveforms.

4. What is the operating frequency for horizontal sync, blanking, and sweep waveforms in a commercial television system?
   a. 15 Hz.
   b. 30 Hz.
   c. 60 Hz.
   d. 525 Hz.
   e. 4.5 MHz.

5. What is the operating frequency for vertical sync, blanking, and sweep waveforms in a commercial television system?
   a. 15 Hz.
   b. 30 Hz.
   c. 60 Hz.
   d. 525 Hz.
   e. 4.5 MHz.

6. Which one of the following statements best represents the difference between a television camera tube and a CRT?
   a. The beam in a camera tube is composed of positively charged particles, while those in a CRT are negatively charged.
   b. A camera tube converts varying amounts of beam current into varying amounts of light intensity.
   c. A camera tube converts varying amounts of light intensity into varying amounts of beam current.

7. A color television system creates color images from these three primary colors:
   a. red, yellow, and blue.
   b. red, green, and blue.
   c. black, white, and gray.
   d. red, white, and blue.

8. Which one of the following features is found in a color television signal but not in a black-and-white signal?
   a. Luminance.
   b. Chrominance.
   c. Superheterodyne.
   d. Blanking pulses.

9. Which one of the following statements is true for commercial television broadcasting?
   a. Video and sound are both broadcast as FM signals.
   b. The video signal is broadcast as AM, the audio as FM.
   c. The video signal is broadcast as FM, the audio as AM.
   d. Video and sound are both broadcast as superheterodyne signals.

10. Which one of the following statements properly describes an important feature of rf electromagnetic waves?
    a. Lower frequencies tend to follow the curvature of the earth.
    b. Lower frequencies travel in straighter lines.
    c. Lower frequencies are better suited for satellite communications.
    d. Higher frequencies are easily reflected by the ionosphere.

Understanding Modern Microcomputers

ABOUT THIS CHAPTER

Computer technology has become a vital part of modern electronics. This relative newcomer stands on a par with the more traditional electronics specialties of audio, communications, industrial, and military electronics. In this chapter, you will learn how computers are organized, how computers respond to different levels of programming, the purpose and organization of computer memories, and the purpose of magnetic tape and disk storage systems.

HOW COMPUTERS DIFFER FROM OTHER KINDS OF MACHINES

Computers are machines. However, computers differ from other, more conventional kinds of machines in several important respects. First, computers are information-processing machines. Computers accept information from the outside world, process it in some meaningful fashion, and send the results back to the outside world. No other kind of machine is designed specifically for that purpose.

Another special feature of computers is that they are basically general-purpose machines. Although some are designed to do a specific information-processing task, most are designed so that they do one kind of task at one moment and then an entirely different kind of task the next moment. A simple home computer, for example, can help you balance your family budget and then play the role of a video game a few moments later. No other kind of machine is so easily adaptable to an endless variety of applications.

A third important feature of computers is their high operating speed. Modern computers can perform millions of simple operations each second. No other kind of machine can do so much in such a short time.

THE BASIC PARTS OF A COMPUTER SYSTEM

Figure 15-1 illustrates the three basic sections of a computer system. Information is fed into the "processing unit"—the "brain" of the system. The processing unit performs the prescribed operations on the original information, then sends out the results.

Figure 15-1.
Basic Sections of a
Computer System

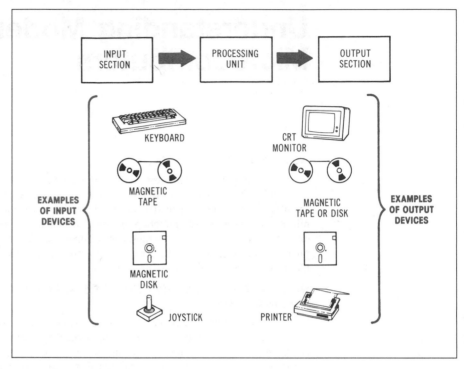

The Input Section

The purpose of the input section is to convert information into electrical codes that suit the requirements of the processing unit.

The processing unit can understand information only in terms of patterns of electrical signals. Most information that is fed into a computer system is not in an electrical form. When you are using a computer to keep track of names and addresses, for example, this information is usually in the form of a printed list. The printed information has to be converted into coded electrical signals that suit the needs of the processing unit. This is the task of the "input section" of a complete computer system.

There are a great many kinds of devices that are used as part of a computer's input section. A typewriter-like keyboard, for instance, translates each key depression into an electrical code. The input section further refines the codes so that they are meaningful to the processing unit.

The joystick on a computerized video game translates the direction and amount of motion of the stick into electrical signals that can be converted into codes that can be understood and manipulated in the processing unit.

Large amounts of information can be fed into the system through the input section from magnetic tape or disks. Furthermore, any event that can be translated into corresponding electrical signals can be fed into a computer through its input section. This includes radar information at busy airports, temperature and radiation information in atomic power plants, voltage and current information in complex electrical machinery, and vital biological signs in medical facilities.

The list of sources of information is virtually unlimited. The only requirement is that we have some technique available for translating the information into the form required by the input section. Once we satisfy the requirements of the input section of a computer system, we know the processing unit can deal with it.

The Output Section

The purpose of the output section is to convert the results from the processing unit into a form that is useful and meaningful for the task at hand.

Just as the information to be processed by a computer is rarely available in a form that is directly compatible with the processing unit, the results of the process are rarely in a form that is directly usable or understandable. The purpose of the "output section" in a computer system is to convert the results of a processing operation into a form that is more useful and meaningful for the task at hand.

One of the most common types of output devices is a CRT. A processing unit does not generate information that can be directly interpreted as video information. But it can produce information, in the form of electrical codes, that an output section can convert into meaningful video information.

A computer's processing unit and output section can be made to generate large volumes of information at high speeds. All of this information can be printed out on sheets of paper or recorded on magnetic tape or disk for study at a later time.

Output sections can be designed to convert the information from the processing unit into just about any imaginable form. The only requirement is the availability of techniques and circuits. For example, an output section can be designed in such a way that it converts information from the processing unit into mechanical motion. This is the principle behind the operation of computer-controlled robots. Other kinds of output sections can convert processor information into complex signals that turn light bulbs on and off, creating the high-tech scoreboards in large sports arenas.

Input and Output Sections Working Together

The nature of the input and output sections of a computer system determine how the system can be used.

You can see that input and output sections are vital to the usefulness of a computer system. No matter what kind of information we are processing, it must be converted into the same kind of electrical coding for the processing unit. Likewise the electrical coding from the processing unit must be converted into useful and meaningful forms. The processing unit in a computer system always works according to the same electrical specifications and codes. It is the nature of the input and output sections that makes one kind of computer application different from another kind of application.

The range of tasks that a computer can perform for you depends on the kinds of "I/O devices" (input and output devices) you use with it. The simplest home computers use just two kinds of I/O sections. One of these I/O sections lets you input information from a keyboard and read it out on a CRT. The other I/O section lets you enter information from magnetic tape or disk and output processed information to the same tape or disk machine.

COMPUTER PROGRAMS

Most processing units can perform fewer than 255 simple operations.

Most of today's processing units can actually perform no more than about 255 different operations. They can do only the simplest kind of arithmetic, count forward or backward, perform some elementary logical operations, and move small amounts of information around within itself. Any one of these operations is too simple to be of any real value.

The relatively few operations that a processing unit can perform, however, can be selected and combined into a sequence of operations that gives us the impression that the unit is performing far more complex and useful operations. Any information-processing task that a computer does can be reduced to sequences of its small family of simple tasks.

The small family of simple operations that a processing unit can perform are selected and combined to do any complex task required of the computer. A program is a set of instructions that are fed to a computer to tell it what it is supposed to do and when it is supposed to do it.

Each of the simple tasks a processing unit can perform on its own is assigned a codes. Feed one of those codes to the processing unit, and the unit performs the task. Feed a long list of those codes to the processing unit, and the unit will perform the operations in the order in which they are presented. A list of codes that instruct the processing unit—tell it what to do and when to do it—is called a "program." An individual who writes computer programs is called a "programmer."

Machine-Coded Programs

Programs that directly use the instruction codes for the processing unit are called "machine-coded programs." Few processing units include instructions for multiplying decimal numbers. The procedure, however, can be carried out by a suitable arrangement of the basic operations. The machine-coded program for multiplying two decimal numbers might include as many as thirty of the simple instructions. *Figure 15-2* shows a set of machine-coded instructions for a processing unit.

Assembler Programming

Assembler programming lets the programmer work with code words, rather than numbers, that represent each of the simple tasks a processing unit can perform.

A computer's processing unit can operate only on the basis of machine-coded programs. These programs, however, are difficult for human programmers to write and understand. For this reason, programmers write such programs in a more humanly understandable form called "assembler programming." When you are using assembler programming, each of the simple tasks that the processing unit can perform is designated as code words that describe the nature of the tasks. Of course you have to learn exactly what the code words mean, but that is far easier than learning and recalling the machine-code versions.

**Figure 15-2.
A Set of Machine-coded
Program Instructions**

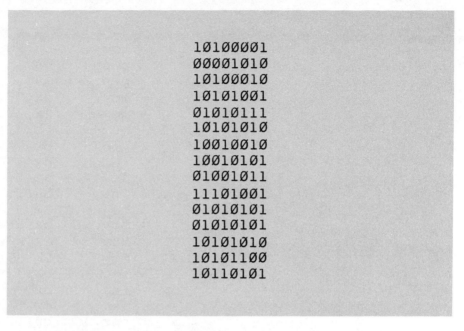

```
10100001
00001010
10100010
10101001
01010111
10101010
10010010
10010101
01001011
11101001
01010101
01010101
10101010
10101100
10110101
```

Assembler programs must
be converted to machine-
coded programs before
they can be executed by
the processing unit. A
program that does this
conversion is called a
compiler.

Once programmers have written assembler programs, they can use
other programs that convert the word codes into the number codes
required for operating the processing unit. A program that converts
assembler codes into machine codes is called a "compiler." *Figure 15-3*
shows a portion of assembler programming.

High-Level Programming Languages

Although assembler programming makes it easier for human
programmers to compose programs, the word codes that are available are
still limited to the family of simple instructions the processing unit can
perform. Whenever assembler programmers want to multiply two decimal
numbers, they still have to write a long list of instructions for doing the job.

As long as a processing unit is good at sorting and moving
information, why not let the computer, itself, help with the job of writing
programs for it? This is certainly possible by means of "high-level
programming languages."

**Figure 15-3.
Portion of an Assembler
Program**

```
              mov      bx,BUFFER
              push     cs
              pop      ds
              mov      cx,0020h
              mov      ax,2020h
     buff1:   mov      [bx],ax
              inc      bx
              inc      bx
              dec      cx
              jnz      buff1

              mov      cx,5FDDh
              mov      ax,0000
     buff2:   mov      [bx],ax
              inc      bx
              inc      bx
              dec      cx
              jnz      buff2
```

High-level programming
languages convert words,
expressions, and symbols
that are easy for pro-
grammers to use into se-
quences of machine-coded
instructions that the
processor unit can use.

There is a wide variety of high-level programming languages.
Some of their names might be familiar to you: BASIC, Pascal, Ada, C,
FORTRAN, and COBOL, to mention a few. High-level programming
languages serve two important purposes. First, like assembler languages,
high-level programming languages replace awkward machine coding with
words, expressions, and symbols that are easier to learn, remember, and
use. What is even more important is that high-level languages include
simple-looking instructions that actually instruct the processing unit to
peform a great many individual operations. Instead of having to specify
several dozen processor operations in order to multiply two numbers, high-
level programming languages let you specify the multiplication of two
numbers in a far simpler and more familiar way—2 * 4, for example. *Figure
15-4* shows a portion of a program written in BASIC.

But just as the central processor cannot work directly with
assembler programs, it cannot work directly with high-level programs.
High-level programs must be converted into the sequences of coded
instructions the processing unit can understand. The conversion process is
done automatically, usually by another program that is built into the
programming language scheme.

**Figure 15-4.
Portion of a BASIC
Program**

```
10 CLS
20 P=1
30 FOR K=1 TO 4:PP(K)=0:PM(K)=100:NEXT K
40 GOSUB 80
50 LOCATE 20,1,0
60 GOSUB 370
70 GOTO 40
80 K$=INKEY$:IF K$="" THEN RETURN
90 DEF SEG=0:POKE 1050,PEEK(1052)
100 IF K$="P" OR K$="p" THEN P=P+1
110 IF P>6 THEN P=1
120 IF LEN(K$)<2 THEN RETURN
130 NP=ASC(RIGHT$(K$,1))
```

COMPUTER MEMORY

You have already seen that a computer system must have a processing unit and some useful I/O devices. Computers must also have some means for storing information. Some kinds of information must be readily available on a permanent basis. Other kinds of information can be stored until it is needed, then erased to make room for storing new information. This is the function of a computer's memory.

Computer systems work with two different classes of information: programs and data. Programs are the instructions a computer requires for doing its prescribed task. Data are information that the computer manipulates in some fashion. An example of computer data is a list of names you enter randomly into the computer. An appropriate program would be one that sorts the names into alphabetical order.

Computers deal with two classes of information: programs and data.

Read-Only and Random-Access Memory

The two kinds of memory built into computers are read-only memory (ROM) and random-access memory (RAM). Programs and data stored in ROM are available on a permanent basis, and as the name implies, the computer can read information from ROM (but cannot write new information to it). One of the main advantages of ROM is that the information remains intact, even when you turn off the power to the computer.

Read-only memory (ROM) holds programs and data on a permanent basis. You cannot change the contents of ROM; the contents are not lost when you turn off the computer.

RAM accepts and holds programs and data on a temporary basis. You can change the contents of ROM; however, the contents are lost when you turn off the computer.

The memory capacity of a computer is specified in terms of the number of bytes of memory it contains. A byte is a segment of eight on/off logic states.

**Figure 15-5.
Computer Memory
Sections**

RAM is intended for the temporary storage of information. Information can be written to RAM, used and manipulated as required, and later erased to make room for other information. Information can remain stored in RAM only as long as power is applied to the system. Whenever you turn off the computer, all information stored in RAM is lost.

Memory Capacity

The memory capacity of a computer is measured in terms of the number of bytes (segments of eight on/off logic states). As illustrated in *Figure 15-5*, a typical small computer might have 16 kilobytes (16 K) of ROM and 48 K of RAM. This means there is 16,000 bytes of built-in ROM information that is always available from the moment you turn on the computer. There is then an additional 16,000 bytes of RAM that is available for your own programming and data.

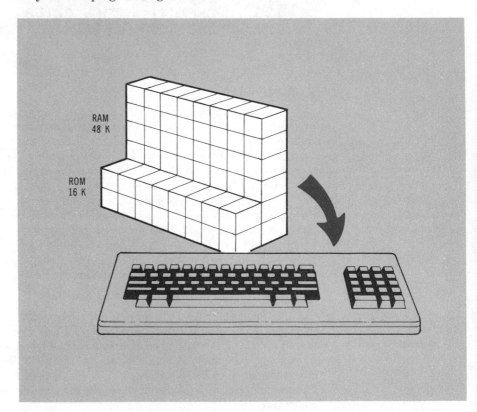

INFORMATION STORAGE AND RETRIEVAL

Large amounts of information can be stored outside the computer and retrieved for later use from magnetic tapes and magnetic disks.

Modern computers must have some means for storing and retrieving large amounts of programming and data. This is done by means of magnetic tape and magnetic disks. Computer information can be stored and retrieved from magnetic tape in the same way audio and video information are recorded and played back on magnetic tape. A magnetic disk works something like a phonograph record, except that the data are magnetically stored and retrieved.

Whether you are working with magnetic tape or disks, you can record the contents of a specified section of computer memory very rapidly. As shown in *Figure 15-6*, this transfer of information from the computer's memory to tape or disk is an output operation.

Information previously stored on tape or disk can be loaded back into specified sections of the computer's RAM when there is a need for using it. This transfer of information from tape or disk to the computer's memory is an input operation.

WHAT HAVE WE LEARNED?

1. Computers are designed to process information. They are highly versatile and operate at high speeds.
2. A processing unit is the "brain" of a computer system.
3. An input section converts information into a form that is meaningful for the processing unit.
4. An output section converts information from the processing unit into a form that is meaningful in the outside world.
5. The processing units within computer systems can do only a few simple tasks according to instructions. However processing units can carry out complex sequences of instructions very rapidly.
6. A computer program is a set of instructions that are fed to the processing unit. A program tells the processing unit exactly what it is supposed to do and when it is supposed to do it.
7. Machine-coded programs are written in the binary form used by the processing unit.
8. Assembler programs use symbols and expressions that are easier for human programmers to use and remember. These programs must be compiled into the machine-coded form before the computer can use them.
9. High-level programs are much easier for human programmers to write, edit, and understand. These programs must be compiled or interpreted before the computer can use them. This is usually done by the computer itself.
10. Computers deal with two kinds of information: programs and data.
11. Read-only memory (ROM) holds information on a permanent basis. A computer cannot change the contents of its ROM.
12. Random-access memory (RAM) holds information only as long as it is required. A computer can change the contents of its RAM.
13. The memory capacity of a computer is specified in terms of the number of bytes of memory. A byte is a segment of eight on/off logic states.

14. Large amounts of computer information can be stored and later retrieved from external magnetic tapes and disks.

KEY WORDS

Assembler programming	Machine-coded programs
Central processing unit (CPU)	Output section
Compiler	Processing unit
Computer program	Program
High-level programming language	Random-access memory (RAM)
Input section	Read-only memory (ROM)
I/O device	

Quiz for Chapter 15

1. The real "brain" of a computer system is called the:
   a. input section.
   b. output section.
   c. processing unit.
   d. memory section.
   e. tape- or disk-storage system.

2. Which one of the following phrases best describes the purpose of a computer's input section?
   a. Serves as a place for storing information on a permanent basis.
   b. Serves as a place for storing information on a temporary basis.
   c. Transforms information from the processing unit into a form that is meaningful in the outside world.
   d. Compiles high-level languages.
   e. Transforms information from the outside world into a form that is meaningful to the processing unit.

3. Which one of the following items is an input device?
   a. Keyboard.
   b. CRT.
   c. Printer.
   d. Program.
   e. High-tech scoreboard.

4. Which one of the following phrases best describes the purpose of a computer's ouput section?
   a. Serves as a place for storing information on a permanent basis.

   b. Serves as a place for storing information on a temporary basis.
   c. Transforms information from the processing unit into a form that is meaningful in the outside world.
   d. Compiles high-level languages.
   e. Transforms information from the outside world into a form that is meaningful to the processing unit.

5. Used in a computer system, a CRT is:
   a. a compiler.
   b. an output device.
   c. the main processing unit.
   d. part of the system's RAM.

6. The processing unit in a computer responds directly to:
   a. machine-coded programs.
   b. assembler programs.
   c. high-level programs.
   d. programs written in BASIC.

7. The main purpose of a compiler is to:
   a. store information on a permanent basis.
   b. store information on a temporary basis.
   c. transform information from the processing unit into a form that is meaningful in the outside world.
   d. transform higher-level programming languages into a form that is meaningful to the processing unit.
   e. transform keyboard information into a form that is meaningful to the processing unit.

8. Which one of the following phrases applies to ROM?

  a. ROM stands for run-off memory.

  b. ROM programming is lost when the computer is turned off.

  c. You should be prepared to load new programming and data into the ROM area of memory.

  d. You can only read information from ROM.

9. Which one of the following phrases applies to RAM?

  a. RAM stands for run-around memory.

  b. RAM programming is lost when the computer is turned off.

  c. You can only read information from RAM.

  d. The memory capacity of a computer is specified in terms of the number of RAM chips it contains.

10. Which one of the following statements best describes the meaning of the expression, 128-K capacity?

  a. The computer contains 128,000 bytes of ROM and RAM.

  b. The computer can handle 128,000 bytes of data per second.

  c. The computer contains the equivalent of 128,000 transistors.

  d. The computer contains the equivalent of 1,024,000 transistors.

Glossary

Ac Amplifier: An amplifier that boosts the voltage or current level of ac signals, generally ignoring dc levels.

Acceleration Anode: An electrode that accelerates electrons as they leave the electron gun in a CRT.

Alternating Current (ac): Current from a power source that changes polarity at a regular rate. Compare with "Direct Current (dc)."

Ammeter: An instrument used for measuring current.

Ampere (A): The basic unit of measurement for current.

Amplifier: An electronic circuit that boosts the voltage or current level of a signal.

Amplitude Modulation (AM): A process of adding information to a radio wave by changing the amplitude of the wave. Compare with "Frequency Modulation (FM)."

Amplitude Modulator: An electronic circuit in a transmitter that modulates the amplitude of the carrier signal with the information to be transmitted.

AND Logic: A principal logic operation characterized by having a true output only when all inputs are true.

Anode: In a semiconductor diode, the anode is the P material that must be made positive with respect to the cathode in order to obtain current flow. In a CRT, the anode is the high-voltage, positively charged terminal that attracts electrons from the electron gun.

Antenna: At a transmitter, the antenna converts radio-frequency signals into electromagnetic waves. At a receiver, the antenna converts electromagnetic waves into radio-frequency signals.

Assembler Program Language: A programming language that refers to the operations directly available from the processing unit but uses expressions that are more meaningful to human programmers. Compare with "High-Level Program Language" and "Machine-Coded Program."

Audio Amplifier: An amplifier designed to operate within the audio-frequency range: 5 Hz to 25,000 Hz.

Audio-Frequency Amplifier: See "Audio Amplifier."

Base: The element in a bipolar transistor that shares pn junctions with the emitter and collector. In many amplifier applications, a small amount of current applied to the emitter-base junction can control a great deal more current between the emitter and collector. In an MOS device, the base is the connection to the silicon foundation material.

Baseline Level: The reference level for a voltage or current waveform. This level is usually the one closest to zero.

Battery: A dc electrical power source composed of two or more chemical cells.

Binary Logic: A system of logic based on just two states—on and off. This is the foundation for all digital technology, including computer technology.

Bipolar Transistor: Transistors that operate according to how a set of two pn junctions are handled. The two possible types of bipolar transistors are npn and pnp. Compare with "Field-Effect Transistor (FET)."

Blanking Pulses: Pulses that are used to turn off the electron beam of a CRT during the horizontal and vertical retrace intervals.

Boolean Notation: A system of notation and rules governing binary logic.

Capacitance: The ability of two conductors, separated by a dielectric, to store an electrical charge. The unit of measurement is the farad (F).

Capacitive Reactance: The opposition to current flow provided by a capacitance. Capacitive reactance is measured in ohms (Ω), and it is inversely proportional to operating frequency and the capacitance value. Compare with "Inductive Reactance."

Capacitor: A device designed to provide a specified amount of capacitance. In the most basic sense, a capacitor is composed of two conductors separated by a dielectric.

Carrier: A radio-frequency oscillation that carries radio and television information. This information is modulated onto the carrier so that it is more efficiently broadcast from a transmitter to a receiver.

Carrier Frequency: The frequency of a carrier signal. Local broadcast stations use different carrier frequencies so that a receiver can select one out of all stations available in the area.

Cathode: In a semiconductor diode, the cathode is the n material that must be made negative with respect to the anode in order to obtain current flow. In a CRT, the cathode is the source of electrons for the electron gun.

Cathode-Ray Tube (CRT): A vacuum tube that accelerates electrons toward a surface that emits visible light wherever the electrons strike it. An ordinary television picture tube is an example of a CRT.

Cell: A device that transforms one form of energy into electrical energy.

Center Frequency: The assigned operating frequency for an FM transmitter. When no information is being sent, the carrier frequency is equal to the center frequency. When sending information, the carrier frequency varies above and below the center frequency.

Central Processor: The portion of a computer that performs the main logical, arithmetic, and data-transfer operations.

Chrominance Signal: The portion of a color television signal that carries the color information.

Closed Circuit: A circuit that provides a complete path for current flow. Compare with "Open Circuit."

CMOS (Complementary MOS): An FET semiconductor technology that is especially useful for applications in digital integrated-circuit devices.

Cold Resistance: The electrical resistance of a heating element that has no current flowing through it. For most heating materials the cold resistance is less than the hot resistance.

Collector: The element in a bipolar transistor that "collects" charges emitted from the emitter and controlled by the base.

Combination Circuit: A circuit that includes components connected both in series and in parallel.

Communications Spectrum: A portion of the radio-frequency, electromagnetic spectrum that is assigned to communications operations.

Compiler: A computer program that converts an assembler or high-level program into the machine-coded form the central processor can understand.

Complete Circuit: A circuit that offers a complete path for current flow. See "Closed Circuit."

Composite Video: A complete video signal that is modulated onto an rf carrier.

Computer Program: A set of instructions that tell the processing unit in a computer exactly what to do and when to do it.

Conductor: A material that passes electrical current very easily. Compare with "Insulator."

Control Grid: The electrode in a CRT that determines the brightness of the light on the screen.

Conventional Current Flow: A theory of current that says current flows from a point of positive potential to a point that is less positive, or negative. Compare with "Electron Current."

Counter emf: A reverse voltage generated by magnetic lines of force cutting the same conductor that carries the current that creates the magnetic lines. (See "Inductor.")

Current: The rate of flow of electrons. Current is measured in amperes.

Dc Amplifier: An amplifier that is capable of amplifying dc levels as well as ac signals. Also called a direct-coupled amplifier.

Deflection Coils: Electromagnets that are fixed to the neck of a cathode-ray tube (CRT) in order to control the position of the electron beam according to the strength and polarity of the magnetic fields.

Deflection Plates: Plates included within the neck of a cathode-ray tube (CRT) in order to control the position of the electron beam according to the strength and polarity of the electrostatic fields.

Demodulator: The section in an FM receiver that converts changes in the frequency of the incoming signal into an audio signal that can be handled by a loudspeaker.

Detector: The section of an AM receiver that converts changes in amplitude of the incoming signal into an audio signal that can be handled by a loudspeaker.

Diaphragm: A metallic disk that vibrates in response to sound waves striking it or vibrates in response to fluctuations in current through an electromagnet. A typical telephone handset uses one diaphragm in the mouthpiece and another in the earpiece.

Dielectric: A type of electrical insulator that is capable of concentrating electrostatic fields. A dielectric material separates the two conductors in a basic capacitor.

Dielectric Constant: A number that indicates the dielectric quality of a material. The larger this value, the better the dielectric effects.

Digital ICs: Integrated-circuit devices that are specifically designed to work with the on-or-off quality of binary operations.

Diode: An electronic device that is designed to pass current in only one direction—from the cathode to the anode.

Direct Current (dc): A description of a source of electrical energy where the polarity does not change. Compare with "Alternating Current (ac)."

Electromagnetic Spectrum: The arrangement of frequencies for all sources of radiation, artificial as well as natural sources.

Electromagnetic Waves: The carriers of radiation, both artificial and natural. Radio and television waves are examples of artificial electromagnetic waves.

Electron Current: A theory of current that says current flows from a point of negative potential to a point that is less negative, or positive. Compare with "Conventional Current Flow."

Electron Gun: The element in a cathode-ray tube (CRT) that produces, controls, and accelerates the electron beam.

Electrons: Tiny, invisible, and negatively charged particles that make up electrical current.

Electrostatic Deflection: A method of deflecting the electron beam in a CRT whereby deflection voltages are applied to vertical and horizontal deflection plates that are built into the neck of the tube.

Emitter: One of the three elements of a bipolar transistor.

Farad (F): The basic unit of measurement for capacitance. A 1-farad capacitor charges at the rate of 1 volt per second when 1 ampere of current is flowing onto its plates.

Federal Communications Commission (FCC): The body of the U.S. government charged with the responsibility of regulating the use of the radio-frequency spectrum.

Field-Effect Transistor (FET): A transistor device that uses electrical charges to control the flow of charged particles through a semiconductor material. Compare with "Bipolar Transistor."

Filament: An electrical heating device. In a light bulb, the heated filament is used for generating visible light. In a cathode-ray tube (CRT), a heated filament increases the number of electrons available from the cathode.

First Detector: An alternative name for the mixer stage in a superheterodyne receiver. (See "Mixer.")

Fixed Resistor: A resistor that has a specified value that cannot be changed by any normal means. Compare with "Variable Resistor."

Focusing Anode: An element in the electron gun of a CRT that sharpens, or narrows, the electron beam.

Forward Bias: A polarity of current or voltage applied to a semiconductor (or vacuum-tube device) that allows the device to conduct current under normal operating conditions. Compare with "Reverse Bias."

Frequency: The number of times a waveform repeats its basic cycle each second. The unit of measurement is the hertz (Hz).

Frequency Modulation (FM): A process of adding information to a radio wave by changing the frequency of the wave. Compare with "Amplitude Modulation (AM)."

Frequency Modulator: The portion of an FM transmitter that modulates the audio signal onto the carrier signal.

Full-Wave Rectifier: A rectifier circuit that converts both half cycles of ac voltage to a pulsating dc voltage. Compare with "Half-Wave Rectifier."

Half-Wave Rectifier: A rectifier circuit that converts only the positive or negative half-cycle of ac voltage to a pulsating dc voltage. Compare with "Full-Wave Rectifier."

Heater: The heating element, or filament, in a vacuum tube. The heater is included as part of the electron gun in a cathode-ray tube (CRT).

Henry (H): The basic unit of measurement for inductance.

Hertz (Hz): The number of complete cycles per second of an ac waveform; the basic unit of measurement for frequency.

High-Level Program Language: A programming language that is written for the benefit of human programmers, using nearly plain-English expressions and ordinary mathematical notation. Examples are Pascal, BASIC, C, and dBASE. Compare with "Assembler Program Language" and "Machine-Coded Program."

Horizontal Deflection: The process of moving the electron beam in a CRT, either electrically or magnetically, in a left-right direction. Compare with "Vertical Deflection."

Hot Resistance: The electrical resistance of a heating element that has full power applied to it. For most heating materials the hot resistance is much greater than the cold resistance.

Image Orthocon: The type of camera tube used for most commercial television stations.

Impedance: Opposition to current flow that combines the opposition from resistance and reactance. The unit of measurement is the ohm (Ω).

Inductive Reactance: The opposition to current flow offered by an inductor. Inductive reactance is proportional to both the operating frequency and the value of the inductor. Reactance is measured in ohms (Ω). Compare with "Capacitive Reactance."

Inductor: A device that uses ac current flowing through a coil wire to create a fluctuating magnetic field that, in turn, generates a counter emf that tends to oppose any change in the current.

Input Converter: A device that converts information from the outside world into a form that is suitable for electronic processing. Compare with "Output Converter."

Insulator: A material or device that does not conduct electricity under normal circumstances. Compare with "Conductor."

Integrated Circuit (IC): A semiconductor device that has a number of diodes, transistors, and resistors etched on the same chip of silicon.

Intermediate Frequency: The fixed frequency that is the result of combining the carrier frequency and local oscillator frequency in a superheterodyne receiver.

I/O Device: A device or electronic circuit that can serve as both an input converter and an output converter.

Kilohertz (kHz): A measure of frequency equal to 1000 Hz—one thousand cycles per second.

Kilovolt (kV): One thousand volts.

Light-Emitting Diode (LED): A semiconductor diode that emits light when current flows from cathode to anode across the pn junction.

Line-of-Sight Transmission: A radio-frequency broadcasting technique that does not allow the transmitted signal to follow the curvature of the earth.

Local Oscillator: In a superheterodyne receiver, the local oscillator produces a frequency that is mixed with the carrier frequency of the selected station to produce an intermediate frequency.

Logic Inverter: A logic circuit that inverts, or reverses, a logic level.

Logic Symbols: Graphic symbols that indicate specific logic functions such as AND, OR, and INVERT.

Luminance Signal: The portion of a television signal that carries the brightness information.

Machine-Coded Program: A computer program that is written in the form directly used by the processing unit. Compare with "Assembler Program Language" and "High-Level Programming Language."

Megahertz (MHz): A measurement of frequency equal to 1,000,000 Hz— one million cycles per second.

Metal-Oxide-Silicon (MOS): An FET semiconductor technology that uses extremely thin layers of metal oxides to insulate gate connections from source, drain, and base connections.

Microampere (μA): A measure of current equal to one-millionth of an ampere.

Microfarad (μF): One-millionth of a farad; a measure of capacitance.

Milliampere (mA): One-thousandth of an ampere; a measure of current.

Mixer: In a superheterodyne receiver, the carrier frequency from the selected station is combined with the signal from the local oscillator in the mixer stage. The result is an intermediate frequency that is the same for all stations. (See "First Detector.")

Modulator: The section in a transmitter that combines the information to be broadcast with the carrier frequency.

Monochrome CRT: A black-and-white picture tube.

Multimeter: A multiple-purpose, multiple-range meter.

NAND Circuit: A logic circuit that inverts the result of an AND logic operation. The output is false only when all inputs are true.

Negative (−): An electrical charge characterized by an excess of electrons. Electron current flow is from a point of negative charge to one that is less negative, or positive. Compare with "Positive."

NOR Circuit: A logic circuit that inverts the result of an OR logic operation. The output is true only when all inputs are false.

Npn Transistor: A bipolar transistor composed of a section of p-type semiconductor sandwiched between sections of n-type semiconductor. Compare with "Pnp Transistor."

Ohm: The basic unit of measurement for resistance—the opposition to current flow.

Ohm's Law: A fundamental law of electricity that says the amount of current flowing in a circuit is equal to the amount of applied voltage divided by the amount of resistance

$$I = \frac{E}{R} \ .$$

Ohmmeter: An instrument that measures electrical resistance.

Open Circuit: A circuit that does not have a complete path for current flow. Compare with "Closed Circuit."

Operational Amplifier: A general-purpose dc amplifier. Most modern operational amplifiers are available as integrated circuits.

OR Logic: A principal logic operation characterized by having a true output when any or all inputs are true.

Oscillator: A circuit that converts dc power into a waveform that is repeated at regular intervals.

Output Converter: A device that converts information from an electronic processing unit into a form that is meaningful in the outside world. Compare with "Input Converter."

Parabolic Antenna: A dish-shaped antenna that receives or broadcasts radio-frequency signals within a very narrow beam. Also called a dish antenna.

Parallel Circuit: An electrical circuit that has more than one path for current flow. Compare with "Series Circuit."

Peak-Inverse Voltage (PIV): The amount of voltage that can be applied to a diode in the reverse direction (negative to anode, positive to cathode) without risking breakdown of the pn junction.

Period: The time, in seconds, required to complete one full cycle of a waveform.

Phosphor: The material on the inner face of a CRT that converts energy from the electron beam into visible light.

Picofarad (pF): One-millionth of a microfarad; a measure of capacitance.

Pn Junction: The electrically charged region where p and n materials come together in a bipolar semiconductor device. A pn junction blocks electron flow in the p-to-n direction but allows electrons to flow in the n-to-p direction.

Pnp Transistor: A bipolar transistor composed of a section of n-type semiconductor sandwiched between sections of p-type semiconductor. Compare with "NPN Transistor."

Positive (+): An electrical charge characterized by a shortage of electrons. Electron current flow is from a point of negative charge to one that is less negative, or positive. Compare with "Negative."

Potentiometer: A type of variable resistor that has terminals fixed at the extreme ends of the resistance and a third terminal connected to a wiper arm that is manually adjusted to select a desired amount of resistance.

Power Supply: An electronic device that converts conventional ac power sources into dc power that is required for operating most transistors and integrated circuits.

Preamplifier: An amplifier that boosts a weak signal level before treating the signal in any other fashion.

Primary Winding: The winding on a transformer that serves as the input connection. Compare with "Secondary Winding."

Processing Device: The portion of an electronic circuit that processes signals before sending them to the output converter.

Processing Unit: The portion of a computer that does the actual manipulation of the digital information. This is often called the central processing unit (CPU).

Program: A set of instructions that tell a computer exactly what to do and when to do it.

Rf Power Amplifier: An amplifier specifically designed to boost the power level of high-frequency signals.

RC Time Constant: The time, in seconds, required for the voltage across a capacitor to change about 63% toward full charge or full discharge. This value can be determined by multiplying the value of the resistor (in ohms) by the value of the capacitor (in farads).

Radio-Frequency Amplifier: An amplifier specifically designed to amplify signals in the radio-frequency (rf) portion of the electromagnetic spectrum.

Radio-Frequency Spectrum: A portion of the electromagnetic spectrum that is devoted to artificial radio and television signals.

Radio-Frequency Transformer: A transformer specifically designed to operate with signals in the radio-frequency (rf) portion of the electromagnetic spectrum.

Random-Access Memory (RAM): A type of computer memory that allows information to be written into and read out of it. Stored information is lost whenever power is removed from the system. Compare with Read-Only Memory.

Raster Scan: A standard scanning pattern for the electron beam in a cathode-ray tube (CRT). This pattern sweeps the beam in a systematic, left-to-right, top-to-bottom path.

Read-Only Memory (ROM): A type of computer memory that has information permanently stored in it. It is possible to read this information, but it is impossible to write new information into it. The information is not affected when power is removed from the system. Compare with "Random-Access Memory."

Receiver: An electronic device designed to intercept radio-frequency signals, amplify the desired frequency, and convert it into a form that is meaningful to the outside world. A radio receiver, for instance, intercepts radio waves and converts them into audio signals. A television receiver converts the information into sound and pictures. Compare with "Transmitter."

Rectangular Waveform: A waveform characterized by a sudden increase to a peak amplitude, some time spent at that amplitude, followed by a sudden decrease to the baseline amplitude. Compare with "Sawtooth Waveform" and "Sinusoidal Waveform."

Rectifier Circuit: A circuit that converts the ac waveform from a power source into a pulsating dc waveform.

Regulated Power Supply: A power supply that provides a specified dc output voltage in spite of normal fluctuations in the circuit and ac power source.

Resistance: The quality of a material to oppose current flow. Resistance is measured in ohms.

Resistor: A device specifically designed to offer resistance to current flow.

Reverse-Bias: A polarity of current or voltage applied to a semiconductor (or vacuum-tube device) that does not permit current flow under normal operating conditions. Compare with "Forward-Bias."

Ripple Voltage: The small amount of ac voltage fluctuation that remains at the output of a dc power supply.

Rosin: A derivative of pine tar that is used in soldering operations to clean the connection while the heat is still being applied.

Sawtooth Waveform: A voltage or current waveform characterized by a slow, constant change in amplitude followed by a rapid change to the baseline level. Compare with "Rectangular Waveform" and "Sinusoidal Waveform."

Schematic Diagram: A graphic representation of a circuit that shows components and their connections in the form of simple, standardized symbols.

Secondary Winding: The winding on a transformer that serves as the output connection. Compare with "Primary Winding."

Semiconductor: (1) A material made from a highly purified insulator, then doped with an impurity to make the material partly conductive. The two types of semiconductor materials are n (electrons carry the charges) and p ("holes" carry the charges). (2) Any device constructed from semiconductor materials.

Series Circuit: A circuit that offers only one path for current flow through all components. Compare with "Parallel Circuit."

Short Circuit: A condition, usually undesirable, where conductors make contact in such a way that current flows through this contact rather than through the intended path.

Silicon-Controlled-Rectifier (SCR): A semiconductor diode that cannot conduct from cathode to anode until it is switched on with a small amount of current at the gate connection. Once the SCR is thus switched on, it can be turned off only by removing or reversing the anode voltage.

Sinusoidal Waveform: A smoothly changing waveform that is naturally generated by a wire rotating within a magnetic field. Energy supplied by power utility companies has this sinusoidal form. Compare with "Rectangular Waveform" and "Sawtooth Waveform."

Skip Effect: A phenomenon that occurs whenever radio waves are reflected from the earth's ionosphere and return to earth at a location far beyond the normal broadcast range.

Solder: A tin-lead alloy that is melted onto an electrical connection to improve the electrical and mechanical properties of that connection.

Step-Down Transformer: A transformer designed so that the voltage at the secondary winding is less than the voltage applied to the primary winding. (See "Step-Up Transformer" and "Voltage Ratio.")

Step-Up Transformer: A transformer designed so that the voltage at the secondary winding is greater than the voltage applied to the primary winding. (See "Step-Down Transformer" and "Voltage Ratio.")

Superheterodyne Receiver: A popular type of receiver that converts the carrier frequency from any selected station to a fixed intermediate frequency. This is done by combining the selected carrier frequency with a signal from a local oscillator in a mixer stage.

Switch: An electrical device that opens and closes circuits. The simplest switches merely turn a circuit on or off. More elaborate switches channel electrical signals through selected pathways.

Synchronizing Pulses: Rectangular waveforms that are used for coordinating electrical events within an electronic system. Synchronizing pulses in a television system are used to keep the scanning pattern on the picture tube in step with the scanning pattern in the camera.

Target: The portion of a television camera that converts an optical image into patterns of electrical charges. These patterns of charges vary the current in an electron beam that sweeps across it.

Temperature Coefficient: A number that indicates what effect a change in temperature has on the electrical resistance of a material. The larger the value, the greater the effect. A positive temperature coefficient indicates that the resistance of the material increases with increasing temperature. A negative temperature coefficient indicates that the resistance of the material decreases with increasing temperature.

Tolerance: The amount that an actual value can vary above and below the specified value without indicating something is wrong. Tolerance is often specified in terms of an allowable percentage of variance.

Transceiver: A piece of communications equipment that both transmits and receives signals.

Transformer: A device that transforms voltage, current, and impedance levels through the interaction of magnetic fields between two coils of wire. Energy is usually applied to the primary winding, and the transformed result is taken from the secondary winding. The amount of transformation is directly related to the turns ratio.

Transmitter: An electronic device that places information onto a radio-frequency carrier signal in preparation for broadcasting it from an antenna. Compare with "Receiver."

Truth Table: A table that portrays the operation of a logic circuit.

TTL (Transistor-Transistor Logic): Bipolar (pn junction) integrated-circuit devices that perform digital logic operations.

Tuner: The section of a radio or television receiver that allows the carrier frequency of the selected broadcast to pass through the system and rejects all other frequencies. The tuner on a television receiver is often called the channel selector.

Turns Ratio: The ratio of the number of turns of wire in the primary winding to the number of turns in the secondary winding of a transformer.

Ultra-High Frequency (UHF): A band of frequencies near the top of the radio-frequency portion of the electromagnetic spectrum.

Variable Resistor: A resistor device that can be easily adjusted across a specified range of resistance values. Compare with "Fixed Resistor."

Vertical Deflection: The process of moving the electron beam in a CRT, either electrically or magnetically, in an up-down direction. Compare with "Horizontal Deflection."

Very-High Frequency (VHF): A band of relatively high frequencies in the radio-frequency portion of the electromagnetic spectrum.

Video Detector: The section in a television receiver that converts the amplitude-modulated (AM) video signal into signals that can be used by the remainder of the video section.

Volt (V): The basic unit of measurement for voltage.

Volt-Ohm-Milliammeter: A type of multimeter that combines the features of a voltmeter (voltage), an ohmmeter (resistance), and a milliammeter (current).

Voltage: The electrical force that causes electrons to flow in a complete circuit. Measured in volts (V).

Voltage Ratio: The ratio of primary-to-secondary voltage of a transformer. (See Step-Down Transformer and Step-Up Transformer.)

Voltmeter: An instrument used for measuring voltage.

Wiring Diagram: A drawing that shows how wiring is routed between electrical devices.

Zener Diode: A semiconductor diode that is specifically designed to operate at a specified voltage level. It is often used as a simple voltage-regulating device.

Index

UNDERSTANDING ELECTRICITY AND ELECTRONICS PRINCIPLES

Answers to Quizzes

Chapter 1

1. a
2. b
3. c
4. b
5. c
6. a
7. a
8. b

Chapter 2

1. e
2. d
3. a
4. c
5. b
6. a
7. a
8. b
9. c
10. d
11. b
12. d

Chapter 3

1. d
2. c
3. a
4. d
5. b
6. a
7. c
8. a

Chapter 4

1. a
2. c
3. c
4. b
5. a
6. a
7. c

Chapter 5

1. a
2. b

3. c
4. a
5. d
6. a
7. d
8. b
9. d
10. c

Chapter 6

1. c
2. c
3. c
4. c
5. a
6. d
7. c
8. c
9. a
10. c
11. d
12. a
13. b
14. d

Chapter 7

1. b
2. d
3. a
4. b
5. c
6. b
7. d
8. a

Chapter 8

1. a
2. b
3. a
4. c
5. a
6. e
7. a
8. b
9. b
10. d

11. a
12. c

Chapter 9

1. c
2. d
3. a
4. b
5. d
6. e
7. f
8. c
9. b
10. b
11. c
12. c

Chapter 10

1. c
2. a
3. d
4. a
5. b
6. d
7. c
8. b
9. b
10. c

Chapter 11

1. b
2. c
3. b
4. a
5. a
6. e
7. e
8. c
9. c
10. a

Chapter 12

1. d
2. b
3. b

4. a
5. c
6. e
7. b
8. c
9. e
10. a

Chapter 13

1. d
2. c
3. e
4. a
5. c
6. e
7. d
8. b
9. c
10. e
11. b

Chapter 14

1. b
2. c
3. c
4. d
5. c
6. c
7. b
8. b
9. b
10. a

Chapter 15

1. c
2. e
3. a
4. c
5. b
6. a
7. d
8. d
9. b
10. a

UNDERSTANDING ELECTRICITY AND ELECTRONICS PRINCIPLES